Environmental Earth Sciences

Series editor

James W. LaMoreaux, Tuscaloosa, USA

For further volumes:
http://www.springer.com/series/8394

Jacques Mudry · François Zwahlen
Catherine Bertrand · James W. LaMoreaux
Editors

H2Karst Research in Limestone Hydrogeology

Springer

Editors
Jacques Mudry
Catherine Bertrand
Faculty of Sciences and Techniques
University of Franche-Comté
Besançon
Doubs
France

James W. LaMoreaux
PELA
Tuscaloosa, AL
USA

François Zwahlen
Faculty of Sciences
Center for Hydrogeology and Geothermics
Neuchâtel
Switzerland

ISBN 978-3-319-06138-2 ISBN 978-3-319-06139-9 (eBook)
DOI 10.1007/978-3-319-06139-9
Springer Cham Heidelberg New York Dordrecht London

Library of Congress Control Number: 2014939063

Printed on acid-free paper

Springer is part of Springer Science+Business Media (www.springer.com)

Acknowledgments

The editors want to warmly thank all the professionals who helped organize the H2Karst Conference in Besançon, France, and who enabled presentation of excellent and relevant articles. These colleagues include Vincent Bichet, Nicolas Bouillier, Martine Buatier, Nicolas Carry, Delphine Charpentier, Didier Convert-Gaubier, Sophie Denimal, Anne-Lise Develle, Charles-Henri Falconnet, Anne-Lise Floch, Boris Gentelet, Céline Guillemot, Romain Ledentu, Christophe Loup, Marie-José Manfredi, Jérémy Marchand, Jacky Mania, Pierre-Antoine Monnier, Jean-Charles Poilvet, Nathalie Puillet, Cristina Ramos, Bruno Regent, Patrick Rosenthal, Jean-Pierre Simonnet, Marc Steinmann, Gisèle Thor, Jean-Daniel Tissot, Pierre Trap, Aurélien Vallet, Sylvie Varey, Chantal Wackenheim, Anne-Véronique Walter-Simonnet (Chrono-Environment, Besançon); Andrea Borghi, Corinne Carraux-Drey, Michiel Pronk (CHYN, Neuchâtel); Marie Belot, Nelly Botella, Francois Gremillard, Jean-Luc Grivet Marc Hamelin, Abderrazzak Kadmiri, Cécile Montigny, Samuel Mourot, Sébastien N'Kodia, Laurence Paillard, Catherine Pellet (Faculty of sciences and Techniques, Besançon), Nathalie Devaux, Aline Nanty, Nathalie Ohl, Bernadette Rassemusse (CNRS, Nancy).

Members of the Scientific Committee are commended for their efforts and for their excellent reviews of the submitted manuscripts. The committee members were:

Bartolomé Andreo Navarro, Malaga, Spain
Iñaki Antiguedad Auzmendi, Leioa, Spain
Bruno Arfib, Marseille, France
Michel Bakalowicz, Montpellier, France
Ralf Benischke, Graz, Austria
Andrea Borghi, Neuchâtel, Switzerland
Catherine Bertrand, Besançon, France
Ognjen Bonacci, Split, Croatia
Pauline Collon-Drouaillet, Nancy, France
Alain Dassargues, Liège, Belgium
Nathalie Dörfliger, Montpellier, France
Wolfgang Dreybrodt, Bremen, Germany
David Drew, Dublin, Ireland
Juan José Durán Valsero, Madrid, Spain
Christophe Emblanch, Avignon, France

Maria-Dolores Fidelibus, Bari, Italy
Derek Ford, Hamilton, Ontario, Canada
Nico Goldscheider, München, Germany
Daniel Hunkeler, Neuchâtel, Switzerland
Pierre-Yves Jeannin, la Chaux-de-Fonds, Switzerland
Hervé Jourde, Montpellier, France
Attila Kovacs, Neuchâtel, Switzerland
Ronald Kozel, Bern-Ittigen, Switzerland
Martin Kralik, Vienna, Austria
James W. LaMoreaux, Tuscaloosa, Alabama, USA
Roland Lastennet, Bordeaux, France
Judit Mádl-Szőnyi, Budapest, Hungary
Barbara Mahler, Austin, Texas, USA
Jacky Mania, Pouilley-les-Vignes, France
Grégoire Mariethoz, Neuchâtel, Switzerland
Jacques Mudry, Besançon, France
Severin Pistre, Montpellier, France
Antonio Pulido-Bosch, Almeria, Spain
Moumtaz Razack, Poitiers, France
Philippe Renard, Neuchâtel, Switzerland
Chris Smart, London, Ontario, Canada
Heinz Surbeck, Neuchâtel, Switzerland
Jean-Pierre Tripet, Bevaix, Switzerland
Sophie Viseur, Marseille, France
Stephen Worthington, Hamilton, Ontario, Canada
Giovanni-Maria Zuppi, Venezia, Italy—*in Memoriam*
François Zwahlen, Neuchâtel, Switzerland

Special acknowledgment is given to the sponsors who made it possible to present the conference on H2Karst:

International Association of Hydrogeologists Karst Commission
Association Internationale des Hydrogéologues - Commission du Karst

International Association of Hydrological Sciences
Association Internationale des Sciences Hydrologiques

Agence de l'eau Rhône-Méditerranée et Corse

Office Fédéral de l'Environnement (OFEV)
Federal Office for the Environment (FOEN)

Centre National de la Recherche Scientifique

Université de Neuchâtel

Université de Franche-Comté

UFR Sciences et Techniques, Besançon

Centre d'Hydrogéologie et de Géothermie de Neuchâtel

Unité Mixte de Recherche 6249 Chrono-environnement

Région de Franche-Comté

Département du Doubs

Ville de Besançon

This book is an important contribution to the research on limestone hydrogeology, providing insight into some of the latest research in karst or limestone terrains. Water resources in karst terrains are being utilized more and for more use; therefore, it is critical for politicians, policy makers, regulators, and the general public to understand the unique nature of their surroundings. Actions, which on the surface appear harmless, can have devastating impacts on foundations, water resources, and the environment. This book helps to better understand these unique situations so as to plan from the beginning how to, or if, to develop an area. If it is decided to develop an area, then it is important to prepare monitoring plans to measure possible impacts and to minimize potential catastophes, if any. The population is continuing to grow and as it does so, people move into more sensitive areas. Contributors to "H2Karst Research in Limestone Hydrogeology" are helping society through their research to better prepare for further development in karst terrains.

The chapters published in this book have been expanded based on additional research and input during discussions at the conference. In view of this, the editors wish to state that any mistakes and omissions or opinions are solely the responsibility of the authors.

Contents

Introduction

Even if karst aquifers cover approximately 15 % of the terrestrial land surface, these water resources play crucial roles, particularly, in many mountainous or agricultural regions. In these regions, karst groundwater resources are more and more utilized for a variety of purposes: drinking water supply, irrigation, power generation. The characteristics of karst systems, namely their relatively high heterogeneity, high hydraulic conductivity, and low efficient storativity, make them vulnerable. Exploitation of these resources by inappropriate land use, however, calls for special surveillance in karst landscapes.

The content of this book is the result of the H2Karst French-Swiss Symposium held in Besançon in September 2011 and reflects the diversity of approaches to karst research at present. The chapters included were selected by the members of the Scientific Committee during the H2Karst Conference.

Establishing the location of karst conduits is difficult and requires the use of indirect methods. One of the trends in research examined in this book is the use of multiple geophysical methods to detect and map karst networks together with anomalies (cavities and palaeo-collapses). Ground Penetrating Radar combined with Electric Resistivity seems to fulfill this task.

Another way to solve the problem of conduit location is to infer a regional karst network, accounting for borehole voids and karst collapses. This simulated network can be input to a 3D coupled continuum-conduit flow model. Then, water balance can be computed within this framework.

To study hydrodynamics and hydrochemistry, the ideal situation is to continuously monitor outlets. This is not possible in very remote places, where access can be sporadic (e.g., every 2 years in Patagonia). Special devices are required on this type of site to obtain measurements between two consecutive visits.

To hydrologically determine recharge altitude of piedmont resources is a challenge for the delineation of protection zones and for the development of strategies for protection policies. Another way to compute water balance in Alpine hydrosystems, which includes karst massif, involves karst sub-basins in distributed hydrological modelling. Hydrodynamics is still a fruitful research field in karst hydrosystems. In moderate climate regions, where recession curves are often interrupted by partial recharge events, hybrid genetic algorithms are necessary to reassemble recession segments.

Hydrograph analysis appears to be a quantitative tool to correlate recession coefficient with aquifer properties: response to diffuse recharge, combination spring-well structure. The hydrodynamic behavior in recession enables parameter estimation both in karst and in connected systems.

In exploitation conditions, the diagnostic plot of drawdown derivative enables the interpretation of pumping tests in heterogeneous conditions under various flow regimes. This method enables separation of conduit flow and matrix flow and allows understanding of karst structure and flow exchanges.

The study of coastal karst deals with freshwater–seawater interaction, accounting for tidal influence, and utilizing monitoring of a hydrodynamic parameter (water level) coupled with physical measurements (temperature, electrical conductivity), and sampling periodically for chemical analyses.

One contemporary issue for protection and management of the quality of karst water is vulnerability assessment. Each method produces different results and requires comparison of one to the other. Comparison of the Slovene and Spanish methods includes sensitivity analyses which demonstrate the importance of parameter scoring.

Another approach to vulnerability is to take into account the behaviour of individual contaminants during their transport and storage. At the field scale, a specific attenuation is observed, due to first-order kinetics. Their impact in the spring is actually determined by intrinsic vulnerability rather than by contaminant-specific properties.

The H2Karst French-Swiss Symposium, held in Besançon in September 2011, is the most recent symposium organized jointly by the Chrono-Environment team in Besançon and the Centre for Hydrogeology and Geothermic of Neuchâtel. From 1971 onwards, after the first conference held in Besançon, hydrogeologists of the Besançon and Neuchâtel teams organize an international event every 4–5 years. In attendance are researchers, students, and practitioners of karst groundwater.

Except for the 1997 meeting, which was organized in the framework of the 12th International Congress of Speleology in la Chaux-de-Fonds, the conferences have been organized in turn in Besançon (1976, 1986, 2001, 2011) and Neuchâtel (1982, 1992, 2006). Over the years, this French-Swiss symposium has taken on a real international dimension with colleagues attending from around 30 countries located on four continents.

Jacques Mudry
Catherine Bertrand
François Zwahlen

Expect the Unexpected! Groundwater Flow in Karstified Carbonate Aquifers

Derek Ford

Abstract Aquifers in carbonate rocks characteristically display triple porosity, the most complex type of groundwater flow system. The complexity arises from antecedent geological conditions, from local geomorphology and climate, and from the patterns of solutional conduits (including accessible caves) and their relationships with any matrix or fracture flow. Much is being learned from study of calcite and other precipitates in such caves. Where not directly accessible, karst aquifers are best approached by study of natural springs, recharge sink points and dye tracing; wells and boreholes are secondary sources. Double continuum and triple porosity designs are most suitable for computer predictive modeling.

Keywords Karst • Triple porosity aquifer • Carbonate solution • Conduit flow

1 Introduction

All carbonate rock formations that display some measure of integrated groundwater flow through interconnected, solutionally enlarged channels (*conduits*) are potentially *triple porosity aquifers* (matrix + fracture + conduit), the most complex type of water supply aquifer. The complexity arises both from antecedent conditioning of the host rock and from current factors of recharge and discharge determined by topographic, geomorphic and climatic controls. This presentation is a personal review of the complexity, based on experience with karst aquifers which began with schoolboy cave exploring in the southwest of England more than sixty years ago. Its focus is upon aquifers fed by meteoric waters where

D. Ford (✉)
School of Geography and Earth Sciences, McMaster University,
Hamilton, ON L8S4K1, Canada
e-mail: dford@mcmaster.ca

J. Mudry et al. (eds.), *H2Karst Research in Limestone Hydrogeology*,
Environmental Earth Sciences, DOI: 10.1007/978-3-319-06139-9_1,

there is little or no confinement beneath impermeable cover strata. These are the so-called *hypergene* or *epigene* aquifers which are the principal underground sources of drinking and agricultural water supplies in the world's carbonate rock terrains. This paper is derived from an invited lecture to the 9th Conference on Limestone Hydrology; it was divided into four parts and illustrated with examples from around the world

2 Antecedent and Physical Geographic Factors that Contribute to the Complexity of Carbonate Aquifers

Sedimentary rocks are classified by their bulk composition (*lime*stone, *sand*stone, *mud*stone, etc.) and subdivided into *facies* differing from each other in details of their particle composition and/or the gross morphology of the accumulations. It is reported that the number of initially distinct facies of limestone and dolomite exceeds that of all other sedimentary rock facies combined (Scholle et al. 1983). It is important to appreciate this great range because facies distinctions can profoundly affect the solubility and effective porosity of the rock matrix, and also help to determine the patterns of fracturing (secondary porosity) that will develop later in it.

Karst and pseudo-karst processes can begin to operate on these facies even as the sediment accumulates—carbonate dunes become cavernous due to mechanical washout beneath case-hardened skins (*syngenesis*), submarine shoals and ramps of calcium carbonate debris exposed by fall of sea level rapidly karstify under the attack of rain waters (*eogenesis)*, intertidal facies become brecciated by dissolution of transient gypsum layers, pauses in shallow submarine accumulation allow some dissolution that produces layers of resistant *hardground*, while slow accumulation at depth may form the soft and much more porous chalks. As an example, U series dating of calcite and aragonite precipitates has shown that some carbonate sand dunes blown up on San Salvador Island (Bahamas) about 125,000 years ago had become strongly cemented and penetrated by solutional caves within 30,000 years or so due to subaerial exposure, an extremely rapid rate of change in geological terms (Mylroie and Carew 1990).

Like other sedimentary rocks, however, the majority of limestones have been transformed more slowly during their subsequent compaction and cementation within larger accumulations in sedimentary basins. In these *mesogenetic* stages the widely differing facies and differing sources of fluid flow further increase the variety of physical and chemical conditions prevailing within the rock pile, and thus the variability of effective permeability and solubility of the consolidating rock (Fig. 1). Effects of dolomitization are always important, while pressure solution may selectively remove as much as 40 % of an already compacted carbonate. Fluids from underlying clastic formations (*transformational* flow—Klimchouk et al. 2000) dissolve further carbonate, leaving vugs and/or emplacements of chert, massive sulphides, etc.

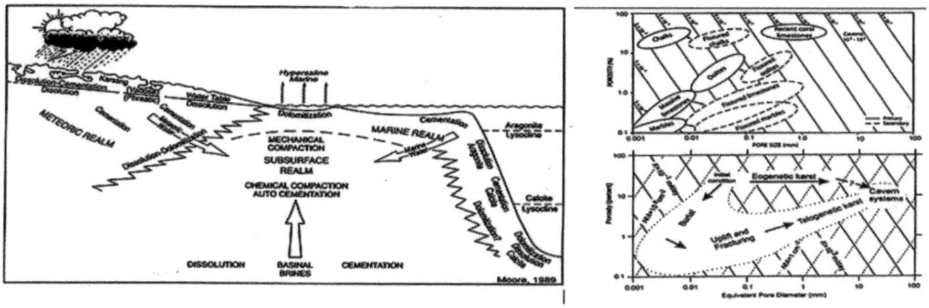

Fig. 1 *Left* Diagenesis of limestone. This model emphasizes the wide range of conditions that may affect a given volume of the rock pile, creating differences in the effective porosity of the matrix and in the patterns of secondary fracturing that may develop subsequently (from Moore 2001). *Right—upper* An idealised relationship of primary and secondary porosity and pore size to the hydraulic conductivity of different limestones following mesogenesis (from Smith et al. 1976)—*lower* Sketch showing how eogenetic karstification may short-cut the mesogenetic processes (from Vacher and Mylroie 2002)

The effects of concurrent or later tectonic folding, faulting and other fracturing are profound, and are considered again in all later parts of the paper. As the carbonate formations (with or without intervening formations of sandstone, shale, coals, etc.) emerge again in response to continental erosion (the *telogenetic* stages) and so become incorporated into meteoric groundwater circulation systems, further transformational flow from e.g. underlying sandstones, may produce *deep-seated karst* reservoirs before the surface-driven meteoric circulation is fully established.

With emergence, the patterns of surface meteoric recharge then become very important. They are determined by stratigraphic, structural and geomorphic controls mingled with meteorological and biogeographical effects. Where large streams can accumulate on other rocks before sinking into carbonates at a geologic contact (*allogenic* recharge) they will usually dominate the conduit drainage in the aquifer. In contrast, precipitation directly onto the carbonates (*autogenic* recharge) produces more dispersed patterns, forming karren fields or doline karsts (Ford and Williams 2007; Gunn (ed.) 2004). Leakage into karst rocks from eroding cover strata such as calcareous shales can create intermediate conditions (*subjacent karst*).

Climatic conditions clearly are very important because they determine the amounts of recharge to aquifers, and the bulk variations of flow through them in both space and time. As volumes of flow increase, so do the quantities of limestone or dolomites being dissolved in most instances. Humidity and higher mean temperatures increase the concentration of carbon dioxide in overlying soils, amplifying the acidity of the water. Climatic factors have received considerable attention in studies of karst landforms because of the strong contrasts between the deeply indented topographies seen in some wet tropical regions with thick carbonate formations (*cockpit karst, tower karst*), and the shallow doline plains found on continental interior platforms in temperate and cool climates (Salomon and Maire

1992). Limestone is abundant everywhere in the rugged Mediterranean region, which has a pronounced summer dry climate; aquifer behaviour and management there and into the semi-arid fringes around it (e.g. Jerusalem) has played a major role in the development of Western cultures since early Biblical times.

3 Meteoric Solutional Patterns where the Conduits Become Big Enough for Human Exploration

From an analytic standpoint modern karst aquifers may be divided into two categories:

(1) those in which at least some of the solutional conduits are large enough to physically explore and map; (2) those which are not, and thus can be approached and probed only as 'black boxes' or 'grey boxes' in Systems Analysis terminology. During the past one hundred years or so several thousands of kilometres of cave conduits have been discovered and mapped by explorers, worldwide and from every climatic and topographic setting (mountain, plateau, plain, coast, submarine) except the high Arctic and the Antarctic. From this broad sample it appears that in any one generation the conduits that develop tend to exhibit branchwork (*dendritic*) or network (*maze*) patterns, with the former predominating below the epikarst at the global scale. They convey turbulent flow, whereas flow in the fractures and rock matrix of an aquifer is mostly laminar. Caves large enough for human beings to enter are only a small proportion of all turbulent flow solution conduits present in most aquifers but their nature and distribution provides insight into the others.

Based on such observations underground, the development of branchwork passage patterns was investigated systematically at McMaster University (1971–1975) by dissolving plaster of paris models to simulate: (1) development of a single conduit (e.g. as in a river meander cut-off cave); (2) multiple stream inputs in one rank (*allogenic recharge*); (3) multiple inputs in multiple ranks (*autogenic recharge*); and (4) restricted input (*river valley under-capture*)—Ewers (1982); see Klimchouk et al. (2000), Ford and Williams (2007, pp. 214–222). Figure 2 illustrates some of the principal findings. Figure 2a sketches the pattern of growth where there is just a single input into a penetrable bedding or fracture. Initially solutional micro-conduits (<1.0 cm in diameter) propagate radially, twisting and turning through the irregularities in the aperture caused by facies or diagenetic factors; sooner or later one or two of them (principal or *p* tubes) advance towards the output boundary, distorting the equipotential pressure field, reducing solvent flow to their competitors (secondary or *s* tubes) and eventually breaking through to the boundary, which permits the much more effective turbulent flow to occur throughout their length. Where there are multiple inputs in one rank, the first one or two *p* tubes to break through will capture their competitors laterally via the failed secondaries located between them, creating simple

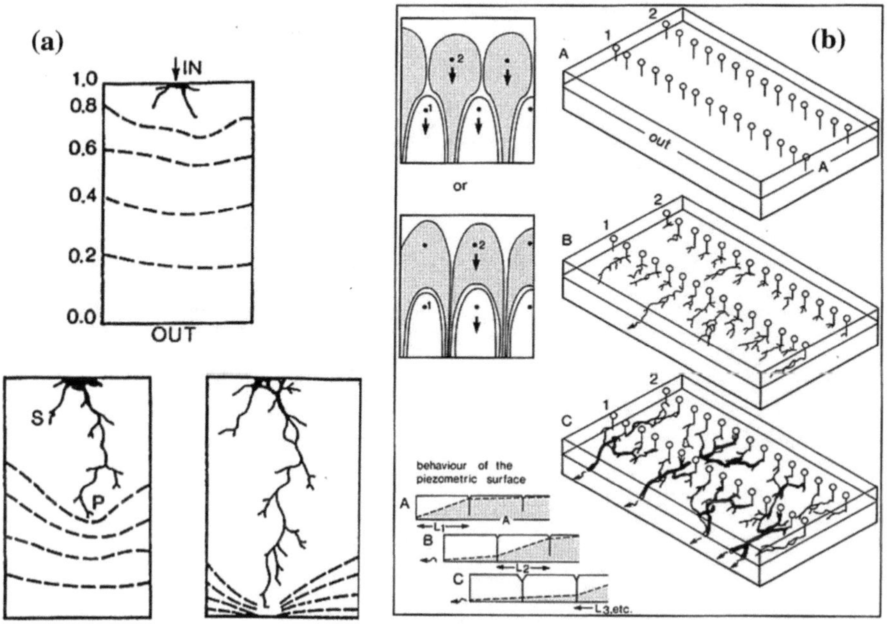

Fig. 2 **a** Model of the development of solution micro-conduits from a single input point in a bedding plane or other penetrable fissure (from Ewers 1982). See the text for details. **b** Model of the progressive development of conduits in a multi-input, multi-rank setting (from Ford and Williams 1989, based on Ewers 1982). See the text for details

dendritic networks. Figure 2b shows the multi-rank or autogenic recharge case that prevails in perhaps the majority of the world's major limestone cave areas. It is seen that the initial flow fields of the rank nearest to the output boundary (the spring line) have the highest hydraulic gradient and obstruct those of the further ranks, with the consequence that earliest development is reduced to lateral competition in that rank in the general case. The *p* tubes that break through first capture the lateral *s* tubes and induce a similar competition in the rank of inputs behind them, as illustrated. This model fits development at scales ranging from small, recently de-glaciated limestone pavements to the largest autogenic doline and cockpit karsts. However, it may be distorted where local recharge conditions give larger volumes of water to selected inputs in rear ranks or where major geologic faults deform the flow fields, etc. Subsequent computer simulations of speleogenesis using thousands of nodes in L × W grids have confirmed these simple principles of conduit propagation that were derived from hardware modeling (see Dreybrodt et al. 2005).

In the years between 1900 and 1970 speleologists paid more attention to the patterns of conduit development in the dimensions of length and depth between inputs and springs in fresh water aquifers. There was much debate on their supposed relationships to the concurrent water tables. Four different propositions

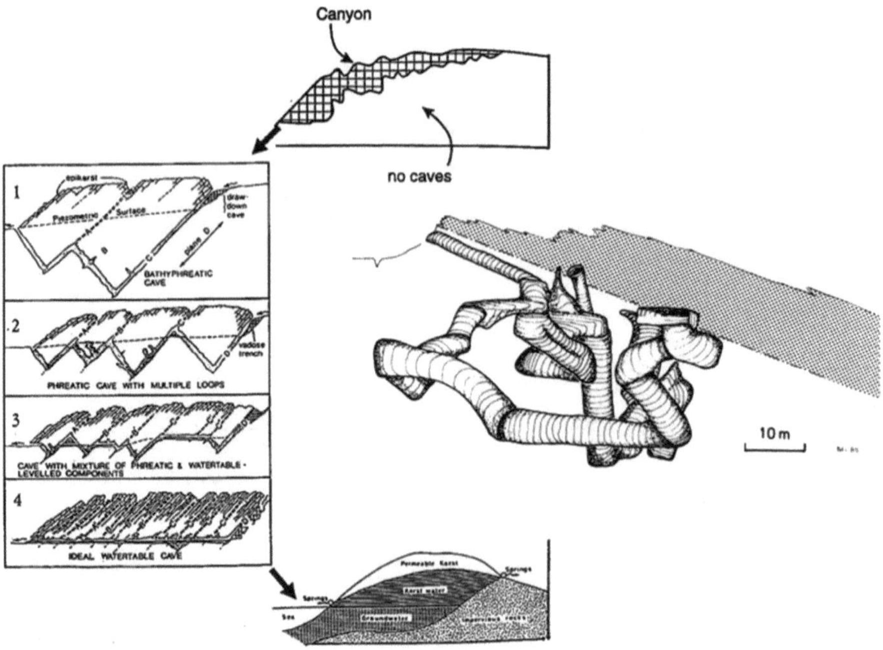

Fig. 3 The extended 'four-state' model of the form of solution caves in the dimensions of length and depth (based on Ford 1971; Ford and Ewers 1978). See the text for details. *Inset*—a 'head-on' depiction of Lake Glomdal Cave, Norway (from Lauritzen et al. 1985)

were advanced—caves develop primarily above the water table, below it, along it propagating from the head, or along it from the output boundary. In 1971 the author showed that all of these possibilities can and do occur in different karst areas, proposing the 'four state' model illustrated in Fig. 3 (Ford 1971; Ford and Ewers 1978; Audra 1994). The fundamental alternatives are: (1) flow through a single 'deep' phreatic loop beneath a shallow vadose zone that has often been created by the conduit volumetric enlargement itself (*drawdown* of the water table); (2) multiple loops, where their upper apices determine the local elevation of the water table; (3) mixtures of such loops with sections graded to an evolving water table; and (4) conduit development along a water table. This variety is determined by the pattern and, especially, the frequency of penetrable fissuration occurring in the host rock. As the frequency increases, so does the likelihood that development will be along or close to a water table that becomes established early in the dissolutional enlargement of the void volume in the aquifer. Mixtures of two or more of these four possibilities may occur due to lithologic or tectonic fracture changes in the rock between the sinks and springs, or during evolution of the caves due to sedimentation or breakdown in them. For example, it is quite common to find major river caves that are well graded to a regional water table in their upstream sections but dive into single or multiple phreatic loops closer to the springs. This is usually due to clastic sediments obstructing early upstream loops, an effect that progresses downstream from the sink points supplying the debris. Some authors have supposed that the 'four-state' model demands that the cave development always progresses from deep loops to water table conduits through successive generations as the elevation of the springs are lowered by allogenic erosion (Audra and Palmer 2012) but this is a misunderstanding: partial progression may occur, as shown for the upstream river cave case noted above, but more often in my experience the successive generations tend to maintain roughly the same amplitudes of looping. Finally, as shown in Fig. 3, the model has two necessary theoretical end members—first, a 'State Zero' where the frequency of penetrable fissures is so low that no breakthrough can occur in the available geologic time, with the consequence that surface erosion processes such as canyon entrenchment destroy the limestone mass instead. This is quite common in many marble outcrops, for example. The second state is the converse, where the frequency of penetrable fissuration (plus matrix effective porosity in some instances) is so high that the ground water flow is dispersed in a multitude of micro-conduits that will not become large enough for human explorers to enter before the limestone itself is removed entirely. This appears to be the case in much of the British chalk, which is a very important aquifer but not an exciting prospect for cave explorers. In broad terms, it was the position adopted by Grund (1903) in some of the earliest writings on the topic.

The dimensions of width and depth across a cavernous aquifer have been neglected by most speleologists in their analyses. Inset into Fig. 3 is the very effective 'head-on' depiction of Lake Glomdal Cave (Norway) by S.-L. Lauritzen. This multi-loop cave is developed in marbles dipping left-to-right across the scene and has been explored and mapped throughout its length by cave divers (Lauritzen et al. 1985).

These basic patterns of conduits propagating and linking together in three dimensions may be complicated by: (a) the contemporaneous formation of mazes of passages in parts of the evolving network due (chiefly) to inter-formational flow from non-karst rocks or to local artesian confinement or floodwater surcharges; (b) *paragenesis* (clastic sediment progressively filling the cave during its enlargement); (c) the introduction of new sinking streams or re-location of old ones as results of climate change or channel avulsions at the surface; and/or by (d) the multi-stage (multi-level) development of interlinked tiers of passages that are caused by changes of spring elevations due to erosion or aggradation. As consequences of these possibilities, the three-dimensional distributions of passages, shafts and large chambers in caves where tens or hundreds of kilometers of accessible passages are now mapped are among the most irregular and complicated patterns known in science. However, they are not at all 'chaotic' in the formal physical sense; on the contrary, they are highly deterministic constructions, reflecting the complexity of carbonate aquifer dissolutional evolution over thousands or millions of years.

4 Speleothems, a Key to Understanding the History of Aquifers Where There are Explorable Caves

Most explored limestone caves contain some stalactites, stalagmites, flowstones or other crystalline wall coatings composed of calcite or aragonite precipitated from waters percolating or flowing in the aquifer, in the phreatic zone as well as above the water table (Hill and Forti 1997). They may cover less than 1 % or as much as 100 % of passage ceilings, walls and floors, giving an observer initial, qualitative insight into the distribution and rates of recharge of waters through the matrix, fractures and inaccessibly constricted conduits in the host rock. This is suggested in Fig. 4, a hypothetical model example where the cave is now entirely in the vadose zone.

Beneath the soil there is well developed epikarst to depths up to a few metres in which the flow and storage conditions can be investigated instrumentally at small scale (e.g. Smart and Friederich 1987). However, between the base of the epikarst and the roof of the cave there is a 'vadose transition zone' that is essentially inaccessible. As suggested by the *downward* arrow in Fig. 4, it is best understood by studying relevant features of the receptor of its flow, the cave. In the figure two deep dolines are seen to reduce the thickness of the transition zone with the result that water passing through it is still acidic when it reaches the cave beneath them, inducing further dissolution with breakdown there. In between these two features the thickness of the zone is greater and rates of recharge and flow are probably much less. In these conditions the water normally becomes chemically saturated within the zone, and will precipitate a portion of its load upon emerging in the cave. Comparatively, rapid dripping tends to produce carrot-shaped stalactites in the roof, with stalagmites and flowstones on the floor beneath them ('1' in Fig. 4); progressively smaller rates of water supply are associated with a regular progression through simple straw stalactites ('2') to straws distorted by helictite extrusions ('3'); finally to very slow precipitation of evaporite aureoles with aragonite ('4') and hydromagnesite ('5'). Surprisingly, little attention has been paid to mapping the distributions of these different types of speleothems in the roofs and walls of caves that are now vadose. It is clear that they record differing conditions in the overlying transition zone, not merely directly above the cave itself but in seepage catchment volumes around it that will widen upwards to the epikarst (Ford and Williams 2007; pp. 291–292). In addition, in the lower levels of multi-stage (level) caves, and in many single stage caves, the modern transition zone will have evolved under antecedent phreatic conditions for a long period before it was drained; study of speleothem distribution in such cases may also give insight into past conditions of phreatic water flow in the rock above and between the conduits that are accessible to explorers.

During the past forty years application of modern lab analytical technology has greatly increased the range and precision of the paleo-hydrological and other environmental information that can be obtained from speleothems (Ford and Williams 2007, 298–320; Fairchild and Baker 2012). The majority of calcite (and the rarer

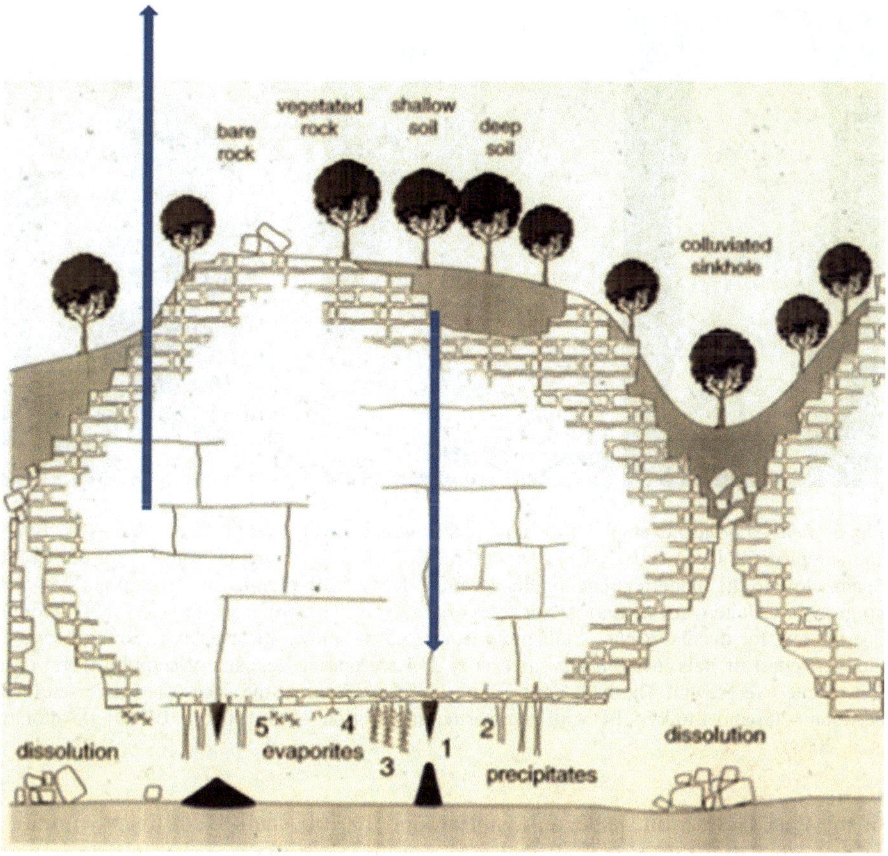

Fig. 4 A depiction of patterns of limestone dissolution and speleothem precipitation in an idealised karst with well-developed epikarst, dolines, a transition zone of varying thickness, and a range of speleothem types; figure adapted from an original drawing by Alain Mangin, with permission. See the text for details

aragonite) speleothems contain sufficient natural U that they can be dated by U series methods over long spans of time. As early as 1973, the author used simple counting of U and Th isotope disintegrations (α particle analysis) to show that stalagmites in a sub-arctic water table cave were greater than 350,000 years in age and thus that the draining of the aquifer must be substantially older. The precision and range of dating by U series methods was much improved by adoption of mass spectrometric measuring techniques after 1987 (Edwards et al. 1986/1987 with corals; Li et al. 1989 with speleothems). Figure 5 gives three examples; *left*, a simple stalagmite sectioned down the growth axis shows many substantial variations in thickness and gray tone in its successive layers, suggesting an eventful history; *centre*, an early example of thermal ionisation dating (TIMS) from the McMaster University lab—a stalagmite from a sub-alpine cave in northern Spain ceased calcite deposition soon after 27,800 years BP due to the onset of the coldest period

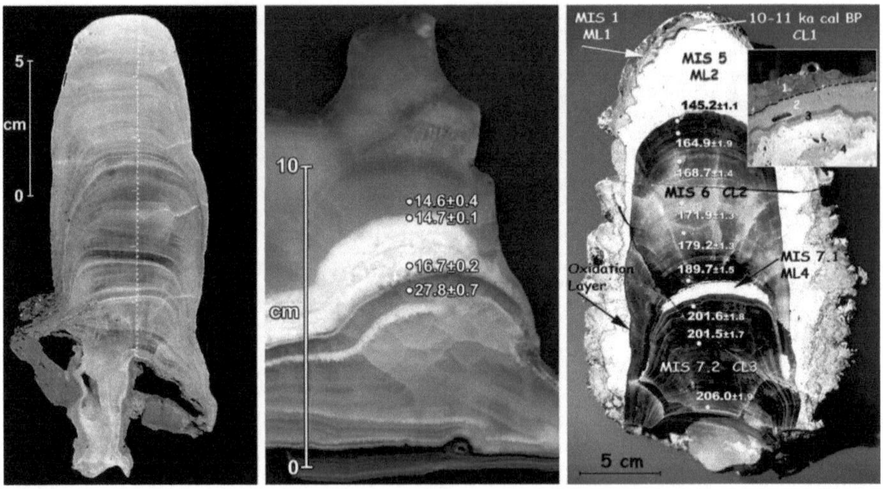

Fig. 5 *Left*—a typical example of the type of stalagmite used for paleoclimate studies, sectioned and sampled for its O and C stable isotope record. *Centre*—a sample from Cueva del Cobre, Spain, close to the modern alpine treeline (from C. Rossi, with permission). It exhibits stop-and-go growth. Calcite (*grey layers*) is being deposited today. The bright white layers are aragonite. See the text for details. *Right*—Stalagmite from 18.5 m below sea level in a drowned cave on the west coast of Italy. ML (*marine layers*) 1, 2, 4 are marine deposits of serpulid worm casts and calcite overgrowth. The hard dark calcite layers grew when the cave was above sea level, between ~206 and 146 kyrs BP with one marine interruption ~200–190 kyrs BP (from Antonioli et al. 2004)

of the Last Glaciation; it began again around 16,700 years BP with precipitation of aragonite that reflected cool and arid conditions at the surface, switching back abruptly to calcite after two thousand years when the climate had become warmer and more humid; *right*, a spectacular example of state-of-the-art induction-coupled plasma dating (ICPMS) of a stalagmite recovered from below modern sea level in Argenterola Cave, Italy, that records the sequence of low sea/vadose calcite and high sea/marine overgrowth events at the site during the last 200,000 years.

The 'ultimate' speleothem dating methods today are those measuring the decay of the uranium to their final end in non-radiogenic lead, U/Pb methods which, in principle, can date the oldest calcites on Earth: its application presents technical difficulties at the present time but they are being overcome in suitable samples, e.g. Polyak et al. (2008) have obtained entrenchment ages as great as 17 million years for the western sector of Grand Canyon, Arizona, by measuring the ages of water table pool calcites in ancient caves now high up in the canyon walls. Lundberg et al. (2000) dated the first karst phase in the giant hypogene caves of the Guadalupe Mountains, New Mexico, to ~90 million years ago, the time of the Laramide Orogeny.

Away from vigorous underground streams cave interior temperatures tend to be very stable and close to the mean annual temperature of the outside environment. As environmental change occurs at the surface, temperatures and other cave parameters

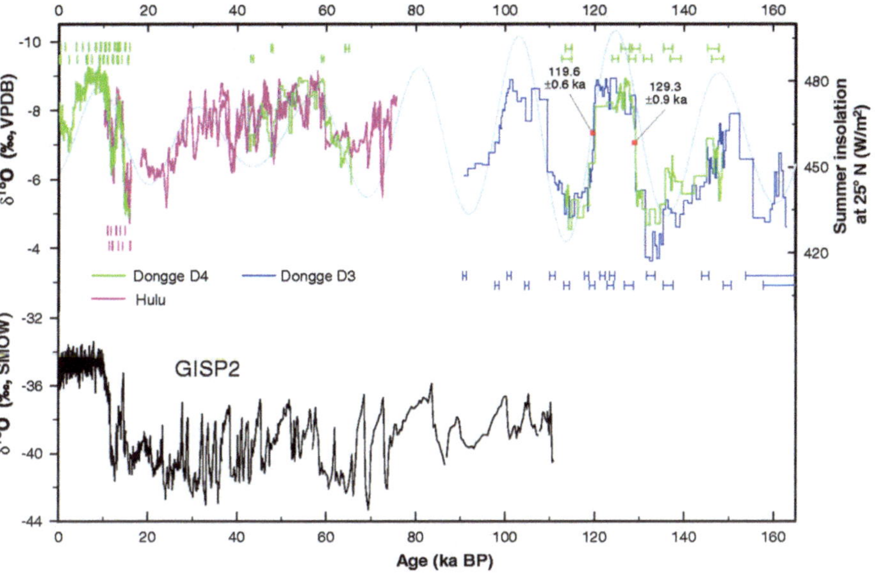

Fig. 6 One of the great successes of oxygen isotope reconstructions of climate change using speleothems. The samples are from Dongge and Hulu caves, 1,300 km apart in eastern China. The cycles seen in their O isotope records are closely similar in age and amplitude, and correlate very well with the computed variation of summer insolation for 25° North. There is also good correspondence with the GISP (Greenland Ice Cap) oxygen isotope record from glacier ice (from Wang et al. 2001)

adjust correspondingly. Most important, the fractionation factors that determine the proportions of the heavier isotopes, ^{18}O and ^{13}C, in speleothem carbonate will change, and also the partition coefficients of common trace metals such as Sr. By the mid-1970s it was reasonably established that ^{16}O:^{18}O and ^{12}C:^{13}C ratios in the carbonate and trapped paleo-waters in U-dated speleothems can track regional and global climate changes through the last three or more glacial cycles (Thompson et al. 1974; indicated by the *upwards* arrow in Fig. 4). With ICP mass spectrometer dating and carbonate micro-sampling techniques, the precision of such climate change records has greatly improved (decadal or better in scale in many instances) and is now attracting much study worldwide; shifts of path and intensity of the east and south Asian monsoons are tracked, records of El Nino and ENSO events sought in Latin American (Fig. 6). The most trustworthy dating of the Quaternary climate cycles during the past 500,000 years is from thermal water calcite crusts in a spring cave in the Nevadan desert (Ludwig et al. 1992).

The percolating waters may transmit many other evidences of environmental changes in the vegetation cover, soil, epikarst and vadose transition zone above a cave. Trace concentrations of fulvic and humic acids and particulate organic matter provide much of the colour in speleothems (van Beynen et al. 2001); and, when luminesced, often record annual or other seasonal detrital flushing events passing through the upper aquifer (Shopov 1987; Genty 1992).

Railsback et al. (1994) reported sequences of calcite-aragonite couplets in a Botswana stalagmite that resulted from seasonal wet-dry cycles in the vadose source zone. Differing thicknesses of such annual accumulations in successive years can give us a measure of the net variations of recharge to the aquifer over many years; for example, Shopov et al. (1994) found seven 'fat' years and seven 'lean' years of deposition in a 27-year long sequence of luminescence couplets precipitated between ~950–1000 AD in a stalagmite from a cave at the forest-to-grassland transition in Iowa.

5 Carbonate Aquifers that We Cannot Explore Ourselves

Most of the volume of the great majority of carbonate aquifers is inaccessible to explorers of human size; so, although it should be presumed that a given example has some proportion of conduit porosity unless there is compelling evidence to the contrary, these aquifers must be treated like those in all other rocks—as 'black' or 'grey' boxes. In North America, where the author has done the majority of his research, the technical approach to carbonate aquifers has been much the same as to all other types of aquifers—to probe the boxes by means of boreholes and observation wells, with a presumption that Darcy flow is predominant. Necessarily, this approach is very costly from the beginning—the holes have to be drilled! In the author's opinion large sums of money have been wasted in the past due to widespread lack of understanding of karstic behaviour among professional hydrogeologists. This review concludes with some personal views on components of 'black box' and 'grey box' exploration methodology.

The first step when approaching an unknown aquifer should be to ***locate and instrument the natural springs***. Detecting them may be difficult where they are masked by aggradation, submergence, etc., but emergence at one or a few points is the norm where there is significant conduit development in an aquifer. Much can be learned from their discharge hydrograph records over one or a few hydrologic years; Mangin (1974/1975) offered a sophisticated analytical classification of aquifer behaviour based on this variable alone. Bakalowicz (1976) and Jakucs (1977) showed how much more can be learned by recording the water temperature, suspended and solute loads in addition; the composition and concentration in many solute loads can be quite precisely determined by calibration with the electrical conductivity (EC) of the water, a variable that can be measured inexpensively and continuously (Krawcyk and Ford 2006).

The next step is to find the natural stream sinks (perennial or seasonal), quantify their discharge and other properties, and trace the flow to the springs by use of dyes or other tracers. The technology has improved greatly in detection limits, breakthrough curve analysis, etc., since the author began in the 1950s (see Dassargues 2000 for a comprehensive review). Smart and Friederich (1987) introduced the practice of dye tracing from the epikarst into underlying caves, amplifying our understanding of groundwater flow and dispersion in the vadose transition zone. Where water wells or observation boreholes are available instead they also are being used to an increasing extent for tracing between wells or from them to natural springs that are more distant. Today, pressure transducers, thermocouples and EC bridges can be linked to flash card recorders inside watertight housings at very little cost (<5 % of 1975 prices?) and operated at remote locations for months or years without servicing. Turbidity metering and measuring organic fluxes by flow-through fluorescence are improving quickly. Sink-to-spring connections can be established and monitored efficiently and inexpensively by dye tracing using modern spectrophoto fluorometers.

To study conditions where there are no accessible sinks and springs, or to investigate underground conditions between them, the first step is usually to apply one or more of the many methods of probing by traversing the surface with ground-penetrating geophysical instrument packages. How useful are they? Every researcher has a favourite method but the author finds that few are successful below shallow epikarst or in detecting deeper cavities "... no one geophysical method has been developed that resolves all the problems of sinkholes and cavities in karst terrain" (Waltham et al. 2004). In the most recent thorough review Kresic (2013, pp. 327–58) broadly concurs with this opinion, stressing that application of two or more complementary methods is better than one and that any interpretation from them must be followed by boreholes. Seismic techniques are the industry standard for deep exploration of bedrock formations and tectonics (e.g. in petroleum exploration) but at more shallow depths the signals tend to badly perturbed by soil cover and moisture content. Measuring the electrical conductivity or resistivity of the ground with arrays of electrodes has been the most widely used method in karst terrains for many years. As a schoolboy the author helped a University of Bristol team use standard resistivity to detect a large cave chamber under the Bristol Downs (England) that had been broken into by lead miners in the 19th Century and then forgotten. The detection was quick and accurate, but the location was already known approximately and the chamber was at shallow depth in very well drained rock, so that the contrast between resistance signals from rock and void were strong and unambiguous. Conditions are rarely this favourable at other sites. Micro-gravimetry has become popular as the cost of the instruments has shrunk and their sensitivity increased. It can achieve some success where the strata are homogeneous and the terrain is quite flat, though there can be problems where the soil thickness varies. Ground penetrating radar is particularly quick and cheap but penetration is usually <10 m and it is said to perform poorly in wet soils; however, the late Bill Wilson found it useful for detection of incipient sinkhole hazards under sands overlying the young limestones in sinkhole-ridden Florida (Wilson 2003, personal communication).

Borehole analysis is a technological field in its own right. The author finds borehole video records to be the most useful of the standard downhole inspection techniques when dealing with karst aquifers. If there are any solutionally enlarged bedding planes or other fractures with turbulent flow in the hole they will usually be detected; if there is no such flow, at the least the video should show the most important levels and intervals for water temperature and solute

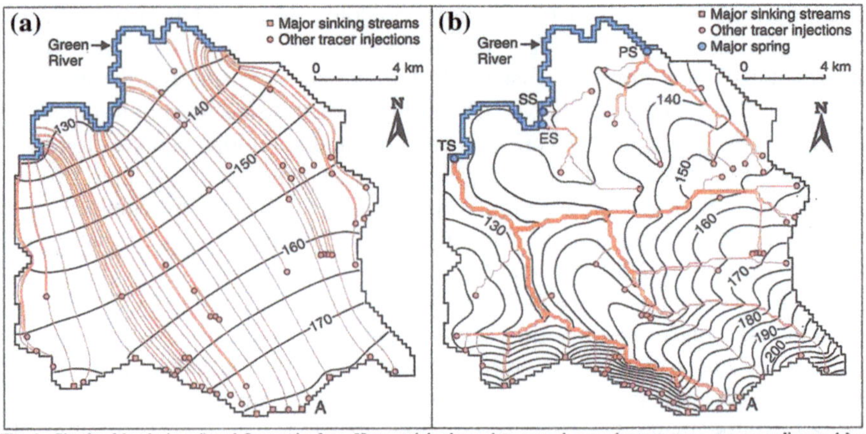

Fig. 6 Simulated heads (m asl) and flow paths from 58 tracer injection points, assuming **a** a homogeneous, porous medium and **b** a channel network. A longitudinal profile of head from point *A* to the *Green River* is shown in Fig. 8b. *TS* Turnhole Spring; *ES* Echo River Spring; *SS* Styx Spring; *PS* Pike Spring

Table 1 Modeling results from the homogeneous porous-medium and karst-aquifer simulations

Parameter	Porous medium	Karst aquifer
Hydraulic conductivity (m/s)	1.1 E-3	4E-5 to 7
Mean absolute error of heads in 48 wells	12.4 m	3.6 m
Mean absolute error of heads in 48 wells and 15 cave streams	-	3.3 m
Turnhole Spring discharge	173 L/s	4500 L/s
Pike Spring discharge	150 L/s	836 L/s
Echo River Spring discharge	201 L/s	359 L/s
Styx Spring discharge	96 L/s	100 L/s
Fraction of total discharge at four springs	10%	95%
Mean tracer destination error	3.97 km	0.00 km

Notes: Mean tracer destination error refers to the difference between the measured and modeled arrival points at the Green River for the 58 tracer tests; spring discharges are taken from the discharge to the river at the appropriate cells

Hydrogeol J (2009) 17: 1665–1678 DOI 10.1007/s10040-009-0489-0

Fig. 7 The comparison between **a** a simple EPM model tracking flow from 58 known dye trace injection points and **b** a karst conduit network model using CAVE, for Mammoth Cave karst area of Kentucky. In the data table below the figure note the extreme differences in estimates of head elevation and spring discharges that occur (from Worthington 2009)

chemical profiling, and for undertaking physical pump or slug (Lugeon) tests between packers. A further important point for consultants is that video imagery is more readily understood and appreciated by the general public, including lawyers and judges in legal proceedings. Dye tracing (between holes or hole-to-spring) is generally more helpful than most inter-hole geophysics in karstic aquifers.

In general hydrogeological practice today a computer model of the groundwater flow is usually considered to be mandatory for any aquifer that is to be managed for water supplies or is threatened by pollution, quarrying, etc. Black box (probabilistic) models focus on recharge events and the responses at the springs and have been developed very well for individual springs since the 1970s, as noted above (e.g. Mangin 1974/1975). Kresic (2013, pp. 435–464) gives a modern review of them. Deterministic (distributed parameter or grey box) numerical models aim to encompass the entire aquifer, not merely the springs, predicting the flow vectors, water table levels, potential yields at different points, etc. These models dominate this field of study today. Opinion is divided on their applicability to karst aquifers as the latter are defined in this paper. "...the heterogeneity of karst aquifers is so severe that it is virtually impossible to acquire sufficient field information to construct a predictive digital model trustworthy enough to allow extrapolation of heads and flow conditions from known to unknown locations, let alone into the future"! (Palmer 2007). The MODFLOW finite difference model, an Equivalent Porous Medium (EPM) model, has become the groundwater industry standard in North America (e.g. Harbaugh 2005). Its codes are based on a single-continuum porous medium concept that assumes an isotropic and homogeneous aquifer with laminar flow (Darcian). It can be modified by stacking layers of differing hydraulic conductivity, e.g. a top layer of high conductivity for the soil and perhaps the epikarst above lower conductivity for the transmission zone. Double porosity (multiple-continuum) models seek to couple matrix + fracture, or matrix + conduit, or fracture + conduit and are currently the most popular in Europe for simulating the observed conditions in most black box or grey box karst aquifers (e.g. Kiraly 1998; Kresic and Stevanovic 2010). The ideal must be triple porosity models (matrix + fracture + conduit); however, the field data are rarely good enough to support them, as Palmer (above) emphasized. One exception, perhaps, is the Mammoth Cave regional aquifer in Kentucky where >650 km of cave passages have been mapped, there are more than 100 observation wells and water wells, many borehole observations and records from a large number of karst springs along the Green River gorge. An accurate water table map has been constructed from them. Using these data, Worthington (2009) used the CAVE program of Sauter and Liedl (2000) to obtain the very good model fit that is illustrated in Fig. 7, where it is compared to a simple particle-tracking EPM that assumes a single homogeneous body of rock. Reimann and Hill (2009) developed a conduit module that has been inserted into basic MODFLOW with some success. Kresic (2013; pp. 479–484) introduces and advocates ... " a true breakthrough in groundwater modelling" 2026 MODFLOW-USG, developed by S. Panday. It is a code newly released to the public domain that (among other applications) allows expansion of the basic MODFLOW package to simulate conduits in three dimensions. It may be that the ideal triple porosity model is at hand; however, it must be remembered that any results can be no better than the quality and quantitative adequacy of the data being entered into the program. The author looks forward to exciting new simulations over the next decade.

References

Antonioli F, Bard E, Potter E-K et al (2004) 215-ka history of sea level oscillations in Argenterola cave speleothems Italy. Global Planet Change 43(1–2):57–78

Audra P (1994) Karsts alpins. Gènese des grands rèseaux souterrains, Karstologia, Mémoires

Audra P, Palmer AN (2012) The vertical dimension of Karst: controls of vertical cave patterns. In: Frumkin A (ed) Karst geomorphology, in treatise on geomorphology, (Shroder JF, General ed), volume 6. Amsterdam, Elsevier pp 186–206

Bakalowicz M (1976) Géochimie des eaux karstiques. Une methode d'etude de l'organisation des ecoulements souterrains. Annales Scientifiques de l' Universite de Besancon 25:49–58

Dassargues A (ed) (2000) Tracers and modelling in hydrology. Publication 262, International Association of Hydrologists

Dreybrodt W, Gabrovsek F, Romanov D (2005) Processes of speleogenesis: a modeling approach. ZRC Publishing, Carsologica, Lubljana 376 pp

Edwards RL, Cheng H, Wasserburg GJ (1986/1987) 238U-234U-230Th-232Th systematic and the precise measurements of time over the past 500,000 years. Earth Planet Sci Lett 81:175–192

Ewers RO (1982) Cavern development in the dimensions of length and breadth. PhD thesis, McMaster University, Hamilton 398 pp

Fairchild LJ, Baker A (2012) Speleothem science. Wiley-Blackwell, Chichester 432 p

Ford DC (1971) Geologic structure and a new explanation of limestone cavern genesis. Cave Res Group Great Brit, Trans 13:81–94

Ford DC, Ewers RO (1978) The development of limestone cave systems in the dimensions of length and depth. Can J Earth Sci 15(11):1783–1798

Ford DC, Williams PW (1989) Karst Geomorphology and Hydrology. Unwin-Hyman, London 601 p

Ford DC, Williams PW (2007) Karst hydrogeology and geomorphology. Wiley, Chichester 562 p

Genty D (1992) Les spéléothemes du tunnel de Godarville (Belgique)—un exemple exceptionnel de concrétionnement moderne. Spéléochronos 4: 3–29

Grund A (1903) Die karsthydrographie: studien aus westbosnien. Geographischen Abhandlungen, Band VII, Heft 3, von A. Penck 7:103–200

Gunn J (ed) (2004) Encyclopedia of caves and karst science. Fitzroy Dearborn, New York, 902 p

Hill CA, Forti P (1997) Cave minerals of the world. National Speleological Society of America, Huntsville, AL 463 p

Harbaugh AW (2005) MODFLOW-2005, the US Geological Survey modular groundwater model. USGS techniques and methods 6-A16

Jakucs L (1977) Morphogenetics of Karst regions: variants of Karst evolution. AkademiaiKiado, Budapest 284 p

Kiraly L (1998) Modelling Karst aquifers by the combined discrete channel and continuum approach. Bulletin d'Hydrogéologie (Neuchatel), 16

Klimchouk AB, Ford DC, Palmer AN, Dreybrodt W (eds) (2000) Speleogenesis; evolution of Karst aquifers. National Speleological Society Press, Huntsville, AL, 527 p

Krawczyk WE, Ford DC (2006) Correlating specific conductivity with total hardness in limestone and dolomite Karst waters. Earth Surf Proc Land 31:221–234

Kresic N (2013) Water in Karst: management vulnerability and restoration. McGraw-Hill, New York 707 p

Kresic N, Stevanovic Z (2010) Groundwater hydrology of springs. Butterworth-Heinemann, Oxford 573 p

Lauritzen S-E, Abbott J, Arnesen R et al (1985) Morphology and hydraulics of an active phreatic conduit. Cave Sci 12(4):139–146

Li W-X, Lundberg J, Dickin AP, Ford DC, Schwarcz HP, McNutt R, Williams D (1989) High precision mass spectrometric dating of speleothem and implications for paleoclimate studies. Nature 339(6225):534–536

Ludwig KR, Simmons KR, Szabo BJ et al (1992) Mass-spectrometric ^{230}Th-^{234}U-^{238}U dating of the Devils Hole calcite vein. Science 258:284–287

Lundberg J, Ford DC, Hill CA (2000) A preliminary U-Pb date on cave spar, big Canyon, Guadalupe mountains, New Mexico, U.S.A. J Cave Karst Stud 62:144–148

Mangin A (1974/1975) Contribution a l'étude hydrodynamique des aquifères karstiques. Annales de Spéléologie 29: 283–332, 495–601; 30: 21–124

Moore CH (2001) Carbonate reservoirs: porosity, evolution and diagenesis in a sequence stratigraphic framework, Developments in Sedimentology 55. Elsevier, Amsterdam 444 p

Mylroie JE, Carew JL (1990) The flank margin model for dissolution cave development in carbonate platforms. Earth Surf Proc Land 15:413–424

Palmer AN (2007) Cave geology. Cave Books Dayton, OH 454 p

Polyak V, Hill CA, Asmerom Y (2008) Age and evolution of the Grand Canyon revealed by U-Pb dating of water table-type speleothems. Science 319:1377–1380

Reimann T, Hill ME (2009) MODFLOW-CFP: a new conduit flow process for MODFLOW-2005. Ground Water. doi:10.1111/j.1745-6584.2009.00561.x

Railsback LB, Brook GA, Chen J, Kalin R, Fleisher CJ (1994) Environmental controls on the petrology of a late Holocene speleothem from Botswana with annual layers of aragonite and calcite. J Sediment Res 64:147–155

Salomon J-N, Maire R (eds) (1992) Karst et evolutions climatiques. Presses Universitaires, Bordeaux, 520 p

Sauter M, Liedl R (2000) Modelling karst aquifer genesis using a coupled continuum- pipe flow model. In: Klimckouk et al. op.cit 212–219

Scholle PA, Bebout DG, Moore CH (eds) (1983) Carbonate depositional environments. Memoir 33, American Association of Petroleum Geologists, Tulsa, OK, 761 p

Shopov YY (1987) Laser luminescent micro-zonal analysis (of calcite speleothems). In: Kiknadze KT (ed) Problems of Karst studies of mountainous countries. Metsniereba, Tbilisi, pp 104–108

Shopov YY, Ford DC, Schwarcz HP (1994) Luminescent microbanding in speleothems: High resolution chronology and paleoclimate. Geology 22(5):407–410

Smart PL, Freiderich H (1987) Water movement and storage in the unsaturated zone of a maturely Karstified carbonate aquifer, Mendip Hills, England. In: Proceedings, Environmental Problems in Karst Terranes and their Solutions. National Water Well Association, Dublin, OH, pp 59–87

Smith DI, Atkinson TC, Drew DP (1976) The hydrology of limestone terrains. In: Ford TD, Cullingford CHD (eds) The science of speleology. Academic Press, London, pp 179–212

Thompson P, Schwarcz HP, Ford DC (1974) Continental pleistocene climatic variations inferred from speleothem age and isotopic data. Science 184(4139):893–895

Vacher L, Mylroie JE (2002) Eogenetic karst from the perspective of an equivalent porous medium. Carbonates and Evaporites 17(2):182–196

van Beynen PE, Bourbonniere R, Ford DC, Schwarcz HP (2001) Causes of colour and fluorescence in speleothems. Chem Geol 175:319–341

Waltham AC, Bell F, Culshaw M (2004) Sinkholes and subsidence: Karst and Cavernous Rocks in engineering and construction. Springer, Berlin 382 p

Wang YJ, Cheng H, Edwards RL (2001) A high resolution absolute dated record late Pleistocene monsoon record from Hulu Cave, China. Science 294:2345–2348

Worthington SRH (2009) Diagnostic hydrogeologic characteristics of a karst aquifer (Kentucky, USA). Hydrogeol J 17:1665–1678

Hydraulic Properties of Karst Groundwater and Its Impacts on Large Structures

Petar Milanović

Abstract A karst environment is a particularly sensitive and risky geological formation for the infrastructure construction from the micro to the mega scale. The hydraulic properties and specific regime of groundwater in karst are, in many cases, the source of catastrophic failures. The most common destructive influence of groundwater is the consequence of: massive turbulent flows; the fast erosion of unconsolidated deposits in caverns and joints; the great kinetic energy of underground flows; propagation of hydraulic pressure at large distances (piston effect); and the enormous hydraulic pressures created in periods of full aquifer saturation, including water-hammer and air-hammer effects due to rapid fluctuation of the water levels. Despite extensive investigations, the destructive impacts are mostly unpredictable in space and time. In many cases these destructive processes take time to become established but final effects appear abruptly, causing considerable damages or failures. The most common consequences of these impacts are subsidence at the urban areas, along the roads and railways, as well as at the bottom of reservoirs; water seepage from reservoirs; break-in of groundwater under high pressure during underground excavation; destruction of surface remediation structures; destruction of tunnel lining; degradation of grout curtains, induced seismicity; decreasing of downstream spring discharges; endangerment of underground species; and the creation of many other unpredictable and unexpected problems. Some dam failures (empty reservoirs) or collapses (entire buildings and factories sinks) were catastrophic. Successful remediation solutions require serious and comprehensive investigations including long period monitoring of groundwater regimes and (in many cases) remedial works during the lifetime of the structure. During construction modifications and adaptations of structures are very common in karst. Persistent, time-consuming and expensive remedial works during the lifetime of the structure are no exception, but, rather, they are the rule.

P. Milanović (✉)
Belgarde, Serbia
e-mail: petar.mi@EUnet.rs

J. Mudry et al. (eds.), *H2Karst Research in Limestone Hydrogeology*,
Environmental Earth Sciences, DOI: 10.1007/978-3-319-06139-9_2,

Keywords Karst • Piston effect • Kinetic energy • Induced seismicity • Dams • Reservoirs • Tunnels • Water seepage

1 Introduction

Karstification is a specific exogenetic process in which the hydrogeological properties of soluble rock masses are changed by solution in vertical and horizontal extension. The principal conditions for karstification are soluble rocks, a large quantity of water and the presence of discontinuities mostly of endogenetic origin. Karstification starts with the chemical action of water (solution) and after flows becomes an important process of turbulent erosion. Both processes (solution and erosion) occurred simultaneously. Mostly the intensity of karstification is not continuous on a geological time scale. Karstification is connected to an emergence phase. Each phase of emergence was in some time followed by the process of karstification of upraised carbonate rocks above the sea level.

As a result of the dynamic neotectonic movement and karst aquifer evolution process, the transformation processes from several karst aquifers into a single aquifer or decomposition of a single aquifer into a few independent aquifers are common. Consequences of these processes bifurcation zones are common in karst, also.

Intensity of karstification increases gradually from initial to the mature stage. In contrast with other natural exodynamic processes, which are mainly limited to the shallow surface zone, the karstification process penetrates into the deeper rock masses mostly along the deep discontinuities. In some cases upraised water (thermal or cold) plays an important role in karstification—hypogene karstification.

In a number of deep and high developed karst aquifers the role of some conduits has been changed during the karst aquifer evolution process. After the aquifer discharge point has been accommodating to the lower erosion base level, the upper channels lose the activity of base flows and stay permanently above the saturated zone. These channels become the cave systems with temporary flows, and, sometimes, are partially filled or completely plugged with cave deposits and sediments.

In Fig. 1 the simplified cross-sections are presented as piezometric lines at different stages of karst aquifer depletion. Concentrated karst flows mean that karst conduits very rarely are continuously inclined. Mostly its horizontal and vertical conduit direction changes frequently. As a consequence, the karst channels in the form of siphons are common and frequent. During dry periods these sections are permanently full of water. Thousands of underground lakes exist in a number of karst aquifers. The large concentration of cave-dwelling aquatic endemic species settled in underground karst lakes all over the world. The exit part of the majority of large karst springs (permanent or temporary) has a form of deep siphons.

As a consequence of the karst evolution process the hydrogeological hierarchy of karst conduits (hydrogeological singularities) is created as one of the

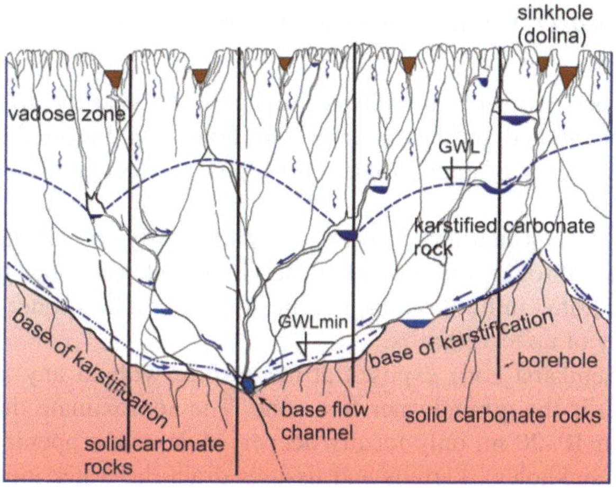

Fig. 1 Simplified cross-section of karst aquifer perpendicular to main flow direction

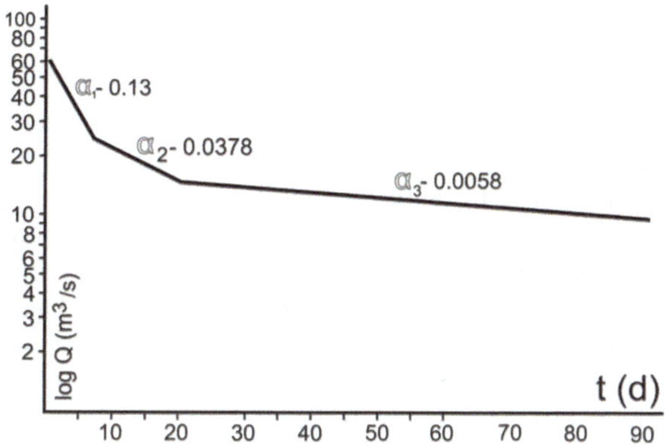

Fig. 2 Ombla Spring recession curve

main properties of a karst aquifer. The top ranking conduits in hydrogeological hierarchy (base flows) represent the base level for the entire karst aquifer. To simplify the explanation of karst aquifer hydrogeological hierarchy, the recession curve of karst springs discharge can be used. The recession curve is a complex function which represents the discharge regime of the aquifer, i.e., the hierarchy of the active karst conduits.

Different inclinations of recession curve correspond to the different discharge coefficients, i.e., to the different hierarchy levels of karst flows. In particular, the case presented in the Fig. 2 part of recession curve with $\alpha_1 - 0.13$ represents the flow in the largest channels, i.e., the top hierarchy channels and flows. The next two sections of recession curve (α_2 and α_3) represent joints enlarged by the dissolution process, and fractures and caverns filled with cave sediments, means secondary and tertiary flows in hierarchy.

1.1 Groundwater Regime of High Developed Karst

The main characteristics of groundwater regime in karst are: extreme fast aquifer saturation, concentrated and fast underground flows, rapid groundwater level fluctuation, kinetic energy of underground flows, piston effect and fast empting of the karst aquifer.

Saturation of karst aquifers is very fast, particularly in bare karst. Saturation occurred mostly through the extremely pervious epikarst zone, but also, in the form of concentrated infiltration through the swallow-holes (ponors), with a swallowing capacity of up to 120 m^3/s.

Mostly the epikarst zone exposed at the surface without any soil cover (bare karstified rock) is the groundwater free zone. The approximate thickness (depth) ranges between 10–20 m, only locally deeper. Most of the openings (karst channels), dolines (sinkholes), karrens and fissures drain the water immediately down to the saturated zone. In many cases 90 % of the water (in spite of the precipitation at 100 mm/10 h) immediately percolate through the epikarst zone and after a few hours reach the aquifer level at a depth of 1,000 or more meters. The negligible part of water can be retarded in the unconsolidated sediments deposited in the karrens, features or bottom of dolines (sinkholes). The thickness of epikarst cover can be from a few meters, up to hundreds of meters. Depending on the cover thickness the epikarst zone can be temporarily saturated, or can be permanently saturated. In some cases the groundwater level is permanently a few meters, or a few tens of meters above the epikarst zone.

Symbolic features in karst hydrogeology are ponors and estavelles, i.e., the features of concentration infiltration. Often ponors are located along the periphery of karst poljes and alongside river beds and river banks. Usually the recharge capacity varies in range of a few m^3/s; however, in a number of cases recharge is a few 10 m^3/s. The capacity of the Ponor Biograd in Nevesinje Polje (Herzegovina) is about 110 m^3/s, and recharge of the Slivlje Ponor in Nikšićko polje (Montenegro) is about 120 m^3/s. From a reservoir integrity point of view, estavelles are particularly dangerous because of double function and ponors cowered (masked) by unconsolidated or low consolidated sediments.

Very fast concentrated underground flows (between 0.5 and 14.0 cm/s, locally up to 50.0 cm/s) with a huge amount of flowing water (10 m^3/s up to the more than 200 m^3/s) have as a consequence large kinetic energy. In some cases the measured exit velocity at the end of the karst channel (spring outlet) was 16 m/s. If the spring outlet is plugged the measured pressure in the channel behind the outlet was between 10 and 11 bars. Another experiment by intensive pumping from syphonic karst spring, in a dry period, shows that the flowing groundwater mass in karst channels has considerable kinetic energy. Explanation of this phenomenon is documented in the graph of the pumping test in Fig. 3.

During the recovery period, due to kinetic energy of underground flow, the water level continued to rise to 130 cm above the starting position of the groundwater level.

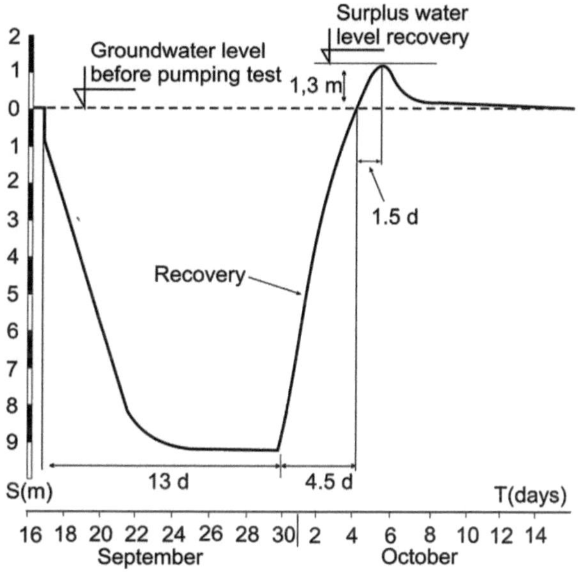

Fig. 3 Graph of pumping test. Jama Spring, Nevesinjsko Polje

As a consequence of fast and turbulent groundwater flows, a very intensive erosion and flushing of clayey deposits from cracks; and cavities frequently occur beneath the dam foundation or along the reservoir banks. It is confirmed that the grouting procedure is an extremely complex task if the water velocity is 10 cm/s or more.

In case of an earthquake, the seismic waves multiply the magnitude of erosion and clay flushing from caves and joints at the dam site. In 1979, after the earthquake in Montenegro (IX° MCS), some consequences were registered at the Gorica dam site (Herzegovina). The distance from the epicenter is about 100 km. In the tailing water, downstream from the dam, the content of eroded clay increased tremendously. Together with thick clay-flow and intensive bubbling the compact clay pieces (0.2–1 kg) were taken out.

Saturated karst aquifers have properties of the hydraulic system under pressure. The piston effect was confirmed by large-scale experiments and analysis of the relationship between the spring discharge and water level in boreholes 4–6 km behind the spring. Between 22 and 78 m^3/s of water was injected into the ponor zone at a distance of 16.5 km away from the spring (Ombla). At the same time, in the ponor zone was injected dye tracer (Fig. 4).

Due to the piston effect the response of the system was much faster than the velocity of underground flow. The spring discharge increased 35 h after water injection. The labeled wave travel time from the same injection point to the spring was nearly four times longer. A very strong correlation between the discharge of the spring and water level in piezometers (r = 0.976) confirms also a very strong hydraulic connection along the saturated karst system.

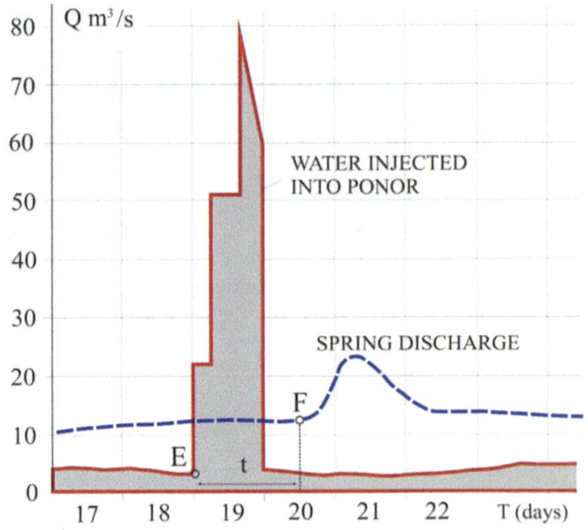

Fig. 4 Comparative graph of water injected into the ponor and the discharge graph in the spring

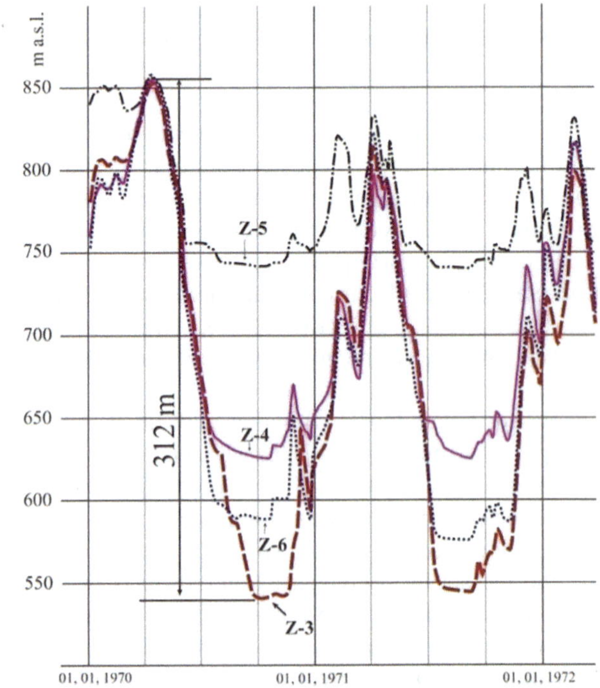

Fig. 5 Groundwater level fluctuations (Nevesinjsko Polje, Herzegovina)

In highly developed karst the groundwater fluctuation is rapid (up to 80 m/24 h) and with huge amplitudes (up to 312 m difference between minimum and maximum) (Fig. 5).

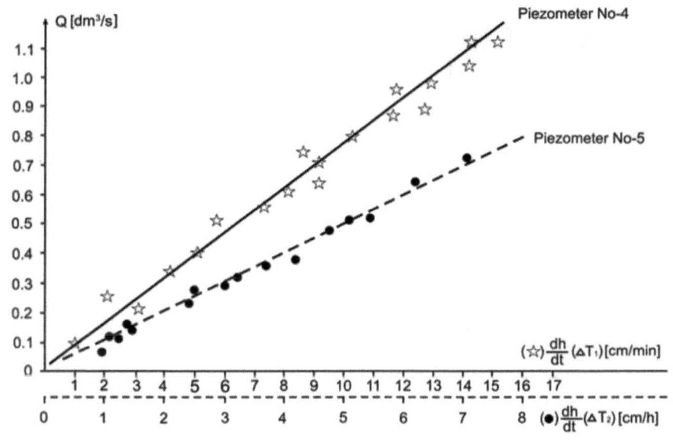

Fig. 6 Relation of air amount (Q) and velocity of water rising in piezometer (Kesić and Kovačina 1978)

The karst aquifer water level graphs are among the most diagnostic sources of hydrogeological information. As a consequence of hydrogeological properties created by karst evolution process and karst aquifer groundwater regime, it is obvious that simple stand pipe piezometers are the best monitoring devices in karst. Double stand pipe piezometers have to be used also if hydrogeological properties require its application only. Systematic use of multi-level piezometers in karst is not justifiable. It is in collision with hydrogeological properties of the nature of the karst aquifer and can be the source of improper design, and, in some cases, may lead to erroneous conclusions.

As a consequence of extremely fast aquifer saturation, the water table rises quickly and removes the air from cavities in the aeration zone. Measurements of air stream velocity from the piezometric pipes confirm good correlation between velocity of water level rising and the amount of air squeezed out. Depending on the volume of the local karst porosity relation between water level rising and the amount of squized air, it is different in separate boreholes; even these boreholes are closely spaced (Fig. 6).

The air stream from some of the piezometric pipes sometimes reaches a velocity of 15 m/s. If karstified rocks are covered with low consolidated sandy-clayey sediments, the fast water table fluctuation provokes strong bottom–up erosion and collapse at the surface of the terrain or at the reservoir bottoms.

One of the most visible aspects in the evolution of karst processes is the existence of well-defined karst horizons. As a consequence, the steady level hydrographs of piezometric boreholes located in such zones are some of the well known characteristics of karst aquifer (Fig. 7).

Discharge of karst springs changes rapidly with enormous difference between minimum and maximum. In some cases, the difference is between 2 and 380 m^3/s (Fig. 8). Discharge of temporary karst springs can vary between 0 and 200 m^3/s.

A karst spring with a pulsation of discharge (intermittent springs) is one of the hydraulic specificities of karst flows.

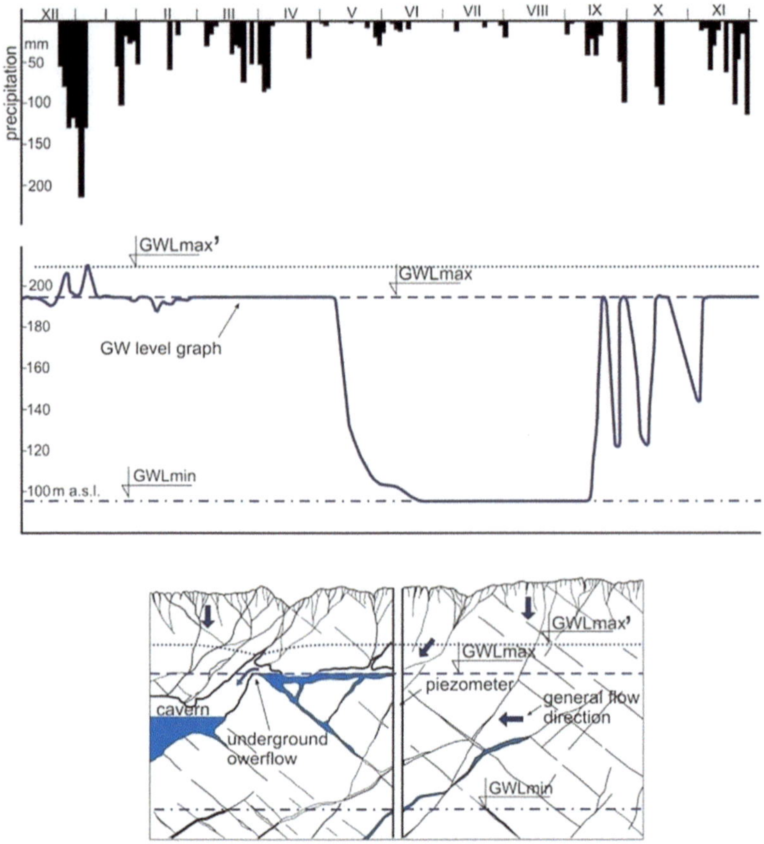

Fig. 7 Steady water level graph indicates presence of karst channel with great flowing capacity

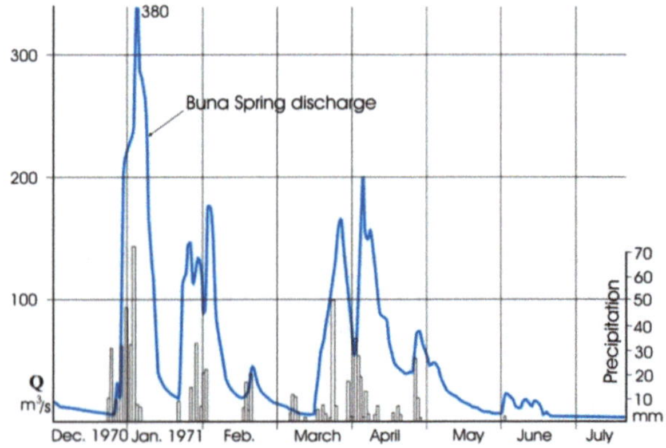

Fig. 8 Buna spring discharge graph (Herzegovina)

2 Permeability of Karstified Rock Masses

Permeability of rock masses is commonly expressed by hydraulic conductivity, Lugeon units or specific permeability. A water pressure test (Lugeon test) still appears to be one of the most commonly applied methods in the framework of hydrogeological investigations. The number of results and long time experience of many specialists raise doubt about the reliability of the Lugeon criteria, particularly for karstified rock masses. Other methods are pumping tests, air pressure tests and tracer tests methods that utilize dye or radioactive tracers.

Analyzing the relationship of the consumption of grouting mixture and the basic permeability (results of WPT), no correlation was established. By comparing the results of WPT (Lugeon test) and grout consumption that were obtained from many examples in karst, it can be concluded that, according to the average permeability, a sound estimation of a possible consumption of grouting mass in the grout curtain cannot be made.

One of the very important reasons for the absence of correlation between Lugeon values and grout consumption is the difference of fluids characteristics. The Newtonian fluid, such as water, used in the case of a Lugeon test, can be characterized by only one parameter—viscosity. The grout mass behaves as a Bingham fluid possessing both viscosity and cohesion.

In spite of its limited practical value, the conclusion of the General report Q58 of the International Commission for Large Dams (Božović 1985) indicates that the Lugeon test remains the main engineering tool in assessing permeability of the dam foundations and in evaluating the achieved efficiency of the grouting treatment.

Lu physically means approximately one joint (aperture 0.2 mm) at each 10 cm of the borehole wall. Results of the WPT test contain also information related to erodability, groutabilty and drainability of rock mass.

If the water pressure test encounters a cavern the double packer method has to be used. The length of the test section, instead of 5 m, should be reduced, i.e., the cavern has to be separated by two packers. That section can be considered as having infinite permeability (unspecified high permeability). Usually it is defined as permeability >100 Lu. A precise location of caverns along the grout curtain route is very important to select the proper water-tightness treatment.

3 Hydrogeological Role of the Base of Karstification

The ***base of karstification*** which represents the transition zone below in which there is no intensive karstification or karstic features are very rare. In general, the base of karstification and the minimum water level coincide. On a regional scale, the base of karstification plays important role from a geological engineering, particularly hydrogeological, point of view.

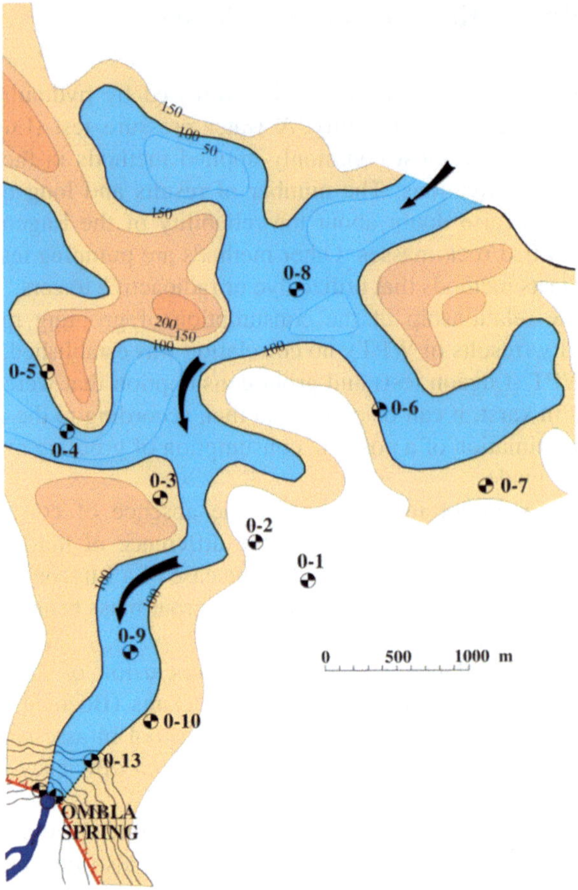

Fig. 9 Contour map for the base of karstification according to results of jointly used electrical profiling and electrical sounding. *Arrows* indicate base flow zone

In many cases the high position of the base of karstification has a role of underground watershed zones. High position of the base of karstification prevents the leakage from Bileća Reservoir (Bosnia and Herzegovina) toward the Bregava River catchment area, as well as the leakage from Piva Reservoir (Montenegro) toward the Tara River erosion base level. The high base of karstification prevents leakage from the Geheyan Reservoir toward the Qingjiang River, China (Ruichun and Fuzhang 2004). A similar situation was found between Cerkničko Polje and Pivka valley in Slovenia (Breznik 1985).

In the opposite deep and narrow base of karstification, a position of concentrated underground flows is indicated. The narrow and deep corridor of the base of karstification, with huge underground flows, was detected by geoelectrical investigations in the catchment area of the Ombla Spring (Fig. 9).

4 Role of Geological Structure on the Groundwater Regime

The number of dams and reservoirs operate successfully because the foundation places were carefully selected on the basis of a superb understanding of a geological structure, its position and groundwater regime. In the opposite, the dam failures attributed to geological causes which mostly occurred due to adverse geological structure and poor hydrogeological data.

The starting point in geological engineering is based on a good and reliable geological data base. A good geological map is needed as a basis from which an analysis of geological structure can be made. Results of such analysis provide the basis for solving any of the engineering problems related to the reservoir watertightness, dam stability, selection of grout curtain geometry, geotechnical treatment of rock mass, tunnel driving, as well as for additional mitigation measures of possible damages during the hydraulic structure operation.

Structural characteristics of the rock mass are, mostly, the consequence of tectonic activities; however, they are also the consequence of exogenetic factors (release discontinuities). Structural relationships play an important role in regional and local directions of groundwater circulation and rock mass quality. Position of impervious lithological units with dominating marly, shaly or clayey component; position of regional and local faults, including mylonite zones; and position of discontinuities (including dip direction, aperture, kind of infilling) govern the groundwater circulation. Wide fault zones, with prevailing mylonite component, are not the proper environment for karstification. In some cases karstification is developed along the secondary joints along the perimeter of the main crushed zone.

Folding structures, especially the compressed anticline cores, are more resistant to karst processes than horizontal and monocline structures. They reduce circulation that is perpendicular to the strike and increase it in the direction of the strike. This conclusion was carried out on the basis of number of case studies in China, Herzegovina, Iran, Tasmania and many other karst regions.

By construction of any large structure the hydrogeological conditions and hydraulic fields in rock mass are drastically changed. Grout curtains become barriers for groundwater filtration and tunnels become huge drains. In both cases the natural groundwater regime is completely changed, sometimes with considerable influence on the surrounding environment.

5 Hypogene Karstification at Dam Sites

An important and quite special problem is the presence of hot or cold up-razed water flows at the areas of some dam sites: Višegrad Dam (BiH), Salman Farsi Dam (Iran), Hammam Grouz Dam (Algeria), Chichik Dam (Uzbekistan), Zhaiziangkou and Pengshui dams, China. According to origin, approximately 10 % of known caves can be classified as hydrothermal or hypogene (Ford and Williams 2007).

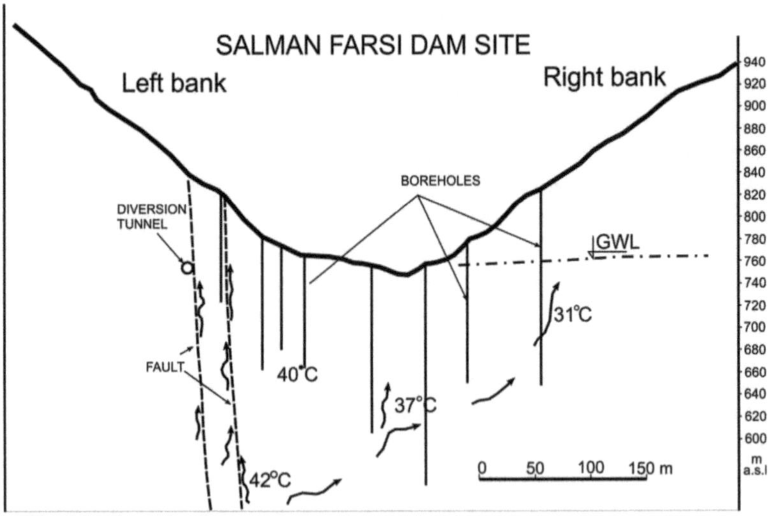

Fig. 10 Deep upward circulation, source for karsification

Mostly the location of the thermal source is very deep with the possibility to be exactly defined as low. However, deep karstification in these cases is obviously the consequence of hypogene flows and, in many cases, requires adaptation of grout curtain route and lower contours. The presence of hot water, particularly if it contains salt, requires particular analysis of possible degradation influence on grout curtains.

Karst features at the Salman Farsi dam site (Iran) are developed as a consequence of hypogene water at depth and simultaneous influence of hypogenic and meteoric water close to the surface (Fig. 10). The number of detected caverns was situated along the grout curtain route at different depths.

In the case of the Salanfe Dam (Switzerland) the leakage from the reservoir has been registered during the first reservoir filling (1953). When the leakage reached its maximum (>1 m^3/s), the new thermal springs (20 l/s) appear in the valley at a distance of 8 km and simultaneously induced earthquakes were registered (Bianccheti et al. 1992).

6 Consequences of Karst Features and Underground Water Regime on Large Structures

As a consequence of the above-mentioned properties, construction of large structures in karst is a risky task. The most common consequences of the listed hydrogeological properties are subsidence at the ground surface as well as at the bottom of reservoirs; water seepage from reservoirs; seepage of highly polluted water from tailings; break-in of groundwater under high pressure during underground excavation and tunnel operation; washing out of clayey/sandy filling from the rock

Fig. 11 Hutovo Reservoir, BiH. Ponors and large cracks created during the first reservoir filling

discontinuities and cavities; destruction of surface protective structures (air-hammer and water-hammer effect); destruction of tunnel lining; progressive degradation of grout curtains by erosion; induced seismicity and specific consequences of natural earthquakes; instability of dam structures due to intensive solution process of evaporates in the foundation rocks; decreasing of downstream spring discharges; endangerment of underground species; other various environmental consequences, and the creation of many other unpredictable and unexpected problems.

6.1 Collapses as a Consequence of Reservoirs Operation

Induced collapses (swallow holes) at the reservoir bottom in karst are spatially independent random events created by reservoir operation. Formation of induced collapses in such a manner is very harmful because their development is unpredictable and practically instantaneous. Collapse at the reservoir bottom is one of the common failures that provoke leakage from the reservoir. Permanent reservoir fluctuations caused the groundwater level fluctuation and washing out of the unconsolidated sediments or cave filled deposits. It provokes collapses in alluvial coverage, as a consequence of ponors in the limestone bedrock and leakage from the reservoir. An inevitable effect is concentrated seepage from the reservoir.

Well-known incidents related to this kind of reservoir leakage are: Lar Reservoir (Iran), May and Cevizli (Turkey), Perdika (Greece), Vrtac and Slano (Montenegro), Angara (Russia), Hutovo (BiH, Fig. 11), Mavrovo (FYUR Macedonia, Fig. 12), Kamskaya (Russia), Huoshipo (China) and North Dike (Florida). In some of them (May, Cevizli, Vrtac) all water quickly leaks out of the reservoir as soon as the rains subside.

Some collapses are extremely voluminous. Among the many collapses which have occurred during the Angara Reservoir operation, the largest collapse has a volume of 7,000 m^3 (Trzhtsinsky and Filipov 1981).

Fig. 12 Mavrovo Reservoir (FYUR Macedonia). Collapses as a consequence of reservoir operation

Fig. 13 Demolished shotcrete at the reservoir bank

The integrity of reservoirs in karst is seriously endangered by water table fluctuation up to the reservoir bottom (presence of estavelles at reservoir bottom). Sudden, abrupt, significant and rapid rise in the water table could cause strong uplift, including air-hammer and water-hammer effects. Another phenomenon, danger for reservoir watertightness, is also related to an increase in the water table. In some piezometers at the Hutovo Reservoir a "breathing" phenomena was registered: air would first rush out (the borehole would "exhale") to be followed with air rushing in (the borehole would "inhale") with each stage lasting for up to 30 min and the cycle could be repeated many times. The air current velocities vary between 0.5 and 1.5 m/s. However, the fastest air current circulation from the aeration pipe can be about 15 m/s (Milanović 2006).

During 2 years of operation of the Hutovo Reservoir (Herzegovina) a number of damages at the reservoir bottom have been recorded: 74 newly created collapses, 1,300 m of fissures (width 2–15 cm), explosion of geomembranes and destruction of shotcrete in a few places (Fig. 13).

After two remedial phases (sealing of collapses and fissures), including construction of aeration tubes to release pressure, the seepage rate was reduced from 3–5 to 1.0 m^3/s (Milanović 2000).

6.2 Water Seepage From Reservoirs

Leakage beneath the dam foundation or through the reservoir banks occurred after the first filling at almost 80 % of analyzed case studies. In such cases the sealing solution requires much patience and perseverance, as well as adequate funds. As a result of a persistent, time-consuming and step-by-step sealing treatment, in some cases the results justify the invested money.

After the sealing treatment, the leakage from Keban Reservoir (Turkey) was reduced from 26 to less than 10 m^3/s; in Camarasa (Spain) from 11.2 to 2.6 m^3/s; in the case of Marun dam site (Iran) the leakage of 10 m^3/s was reduced to the negligible amount; in the example of the Great Falls (USA) from 9.5 to 0.2 m^3/s; in Canelles Reservoir (Spain) from 8 m^3/s to a negligible amount; in Mavrovo reservoir (FYUR Macedonia) a leakage of 9.5 m^3/s was considerably reduced; Buško Blato (BiH) from 40 m^3/s in natural conditions to 5 m^3/s; and in the case of Hutovo Reservoir (BiH) from 10 m^3/s in natural conditions to approximately 1 m^3/s.

In the case of El Cajon Dam (Honduras), the sealing treatment was successful in spite of very deep karst flow. The large karst cave situated 176 m below the gallery had to be grouted under an extremely high head. To reach deep cavities at this dam site the maximum borehole length of 250 m was applied (H. Kreuzer, Personal Communication).

However, in some cases, despite an extensive investigation program and sealing treatment, the results were inadequate to justify the time and money because the hydrogeological conditions, as a consequence of karstification, were too complex. After the first filling of Hails Bar reservoir in USA the leakage was enormous, 54 m^3/s. Corrective treatments started in January 1919. The dam was acquired by Tennessee Valley Authority in 1939. Unable to overcome the continual problems with foundation and leakage, TVA replaced Hales Bar in 1968 with Nickajack Dam 10.3 km downstream. Montejaque Dam (Spain) is abandoned because of unacceptable high leakage—4 m^3/s (Fig. 14).

In the case of Vrtac Reservoir (Montenegro), leakage reached 25 m^3/s; Višegrad Dam (Bosnia), 14 m^3/s; Iliki (Greece), 13 m^3/s; Salakovac (BiH), >10 m^3/s; Lar Dam (Iran), 10.8 m^3/s; Ataturk (Turkey), >11 m^3/s; Samanalawewa (Sri Lanka), 2 m^3/s; Perdikas (Greece); Mosul (Iraq) and May (Turkey) the water-tightness treatment was unsuccessful.

In a karstic environment with a highly random distribution of karstic features some uncertainties always remain. In karst *modifications during grout curtain or tunnel construction are therefore the rule and not the exception.* Re-design of anti-seepage structures has been applied in the case of Salman Farsi, Seimareh and Kavar dams in Iran, as well as in the cases of Frimen Dam (USA) and Sklope Dam

Fig. 14 Abandoned Montejaque Dam, Spain

Figure. 15 Lar Dam, Iran. One of the collapses induced by reservoir operation

(Croatia). In many other cases the number of local modifications (re-routing, extensions, increasing the number of rows, increasing of curtain depth) were applied.

Problems related to the progressive erosion of clayey/sandy deposits from faults, joints and caverns have as a consequence increased leakage from reservoirs after many years of operation: Mavrovo Reservoir (FYUR Macedonia), after 25 years of operation; and Hammam Grouz (Algeria), after 17 years of operation. In the case of Višegrad Dam (Bosnia), during 20 years of operation, leakage gradually increased from 1.2 to 14 m^3/s (roughly 0.7 $m^3/s/year$).

Deep and fast flows—doubtful treatment. In the case of some reservoirs or tunnels, very deep underground flows had caused partial or total failure of the project. In the case of Lar Dam (Iran) the leakage problem is one of the most complicated and extremely difficult problems to be solved using the common geotechnology (Fig. 15).

The deepest karst leakage flows (10.8 m^3/s) from the Lar reservoir have been discovered down to 430 m below the riverbed. However, fractured karstic limestone is extended to at least 700 m below the river bed. At 210 m below the river bed a large cavern was found and plugged (23 m high and 67 m wide, i.e., more than 90,000 m^3). Few other caverns have been discovered at the depth of 250 m down to 430 m below the river bed. In spite of the extensive grouting and cavern filling, the reservoir losses are still almost the same as before treatment (Djalaly 1988).

The main grout curtain at Ataturk dam site is 175–300 m deep below the river level. The grout curtain bottom is high above the base of karstification. The total leakage through the dam foundation in May 1996 was 11–14 m^3/s for reservoir level 6 m below the normal storage level (Riemer et al. 1997).

In the case of Špilje Dam (FYR Macedonia) the main portion of water losses (2 m^3/s) occurs through the deep karst conduits. The deepest conduits are indicated more than 250 below the dam foundation.

In another case, Višegrad Dam (Bosnia and Herzegovina), the main leakage conduits are encountered at the depth of more than 130 m below the dam foundation (14 m^3/s).

However, in some cases the deep sealing works are successful. There are some examples of very deep but successful grout curtains. In those cases the velocity and capacity of underground flows was limited, or filtration occurs through the tight joints. It was the technical prerequisite for successful deep grouting.

In the case of Berke Dam (Turkey) the uppermost grouting gallery is at the elevation of 346 m and bottom of deepest grout curtain contour line is at elevation of minus 50 m. It is one of the most complicated grout curtains in the world (Altug and Saticioglu 2000).

6.3 Underground Treatment

To prevent underground filtration beneath the dams and leakage from reservoirs in karstified rocks the commonly applied underground geotechnical measures are: grout curtain, positive cut-off, diaphragm wall, bath-tub structure and karst cavern sealing.

6.3.1 Grout Curtains and Cavern Sealing

To define extension and lower contour of the suspended curtain, the rock mass permeability is common criteria. The question of target permeability in karst is still open. The prevailing opinion is that permeability less than 5 Lu indicates sufficiently impervious and practically ungroutable rock.

To construct a successful grout curtain in karst the closely spaced grouting galleries are required (less than 30 m between galleries). The following grouting materials are commonly applied: cement with different additives, cement/clay, clay/cement, polyurethane foam, asphalt or hot bitumen and different types of cement mortar.

Almost always grout curtains in karst are combined with plugging of concentrated underground flows or caverns. Surfaces of some important grout curtains in karst are presented below:

Ataturk (Turkey)	1,200,000 m^2, length 5.5 km, depth up to 300 m
El Cajon (Honduras)	610,000 m^2
Limmernboden (Swiss)	544,740 m^2
Berke (Turkey)	533,000 m^2, depth up to 235 m
Buško Blato (Bosnia)	475,000 m^2
Dokan (Iraq)	471,000 m^2
Khao Laem Dam (Thailand)	437,000 m^2
Slano (Montenegro)	404,224 m^2, length 7.011 m
Keban (Turkey)	338,000 m^2
Salman Farsi (Iran)	261,000 m^2

Extreme inhomogeneity of karst porosity leads to great variability of grout mix consumption along the grouting hole. For instance, in the case of Salman Farsi (Iran) the average consumption of grout curtain is 79 kg/m; however, the consumption range varies between 6 and 234.048 kg/m. A consumption rate less than 100 kg/m in karst is rare (only 17 % of analyzed cases). The consumption rates vary mostly between 100 and 600 kg/m (in almost 70 %).

The term ***"karst cavern sealing"*** along the grout curtain or at reservoir banks means: the geotechnical operation needed to block the groundwater circulation along any karst singularity (channels or caverns) which cannot be treated by applying a conventional grouting technology. This term also includes the treatment of potential leakage paths along the heavily tectonized and wide zones filled with mylonite or re-deposited clay/sandy material.

Plugging of large and accessible cavities along the grout curtain route requires different technologies, including speleological investigations, excavation of access shafts and adits, large diameter boreholes, and, sometimes, re-routing the weak places at the grout curtain. If caverns are situated below the water table the first step is definition of cavern size. If a cavern is detected by a borehole, one of the best methods to establish cavern contours is echo logging. The cavern walls are recorded by horizontally and vertically oriented ultrasonic transducer. Good results were also provided by borehole TV camera.

For access and filling of the cavern by aggregate, large diameter boreholes should be used. The next step is consolidation grouting of the aggregate and finally grouting the contact between the concrete plug and cavern walls.

Volume of some large caverns at different dam sites is presented below:

Keban (Turkey)	600,000 m^3 Petek Cavern 150,000 m^3 Crab Cavern
Salman Farsi (Iran)	>150,000 m^3 Golshan's Cavern
Lar (Iran)	90,000 m^3
Pueblo Viejo (Guatemala)	60,000 m^3
Sklope (Croatia)	25,000 m^3
Canelles (France)	10,000 m^3
Slano (Montenegro)	6,000 m^3
Dokan (Iraq)	5,000 m^3
El Cajon (Honduras)	5,000 m^3

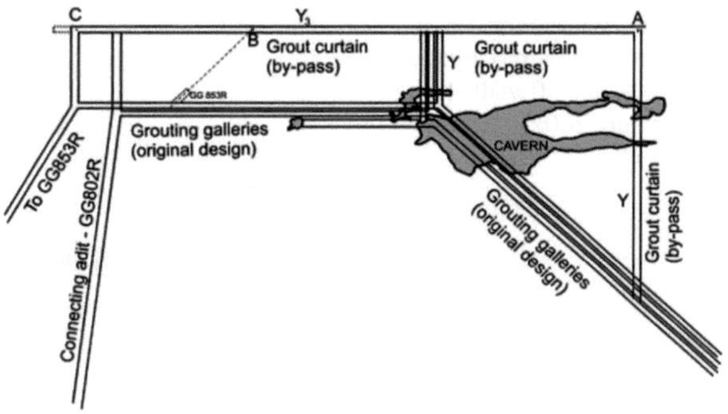

Fig. 16 Salman Farsi Dam (Iran). Grout curtain by-pass around a large cavern

In the case of Keban Dam the Crab Cavern (below the dam foundation and below the groundwater level) was filled with concrete and injected solids. The large Petek Cavern (at the left dam site bank, 150 m from the dam body) has been filled through the shaft 2.5 m in diameter and 13 large diameter boreholes. About 605,000 m^3 of limestone blocks, gravel, sand, and clay were thrown into the cavern. Leakage rate from the Keban Reservoir decreased from 26 to 9 m^3/s.

The problem of Golshan's cavern (Salman Farsi Dam, Iran) has been solved by re-designing the grout curtain in the right abutment to by-pass the cavern from the upstream side (Fig. 16). To plug six large caverns along the grout curtain route at the same dam site 3.125 m^3 of the SCC (Self Compacting Concrete) was used.

One of the largest plugs was constructed as part of the Wulichong underground dam in China. The plug size is: 33.46 m high, 13.9 m wide, and 2–10 m thick, and volume 4,811 m^3 of concrete (Zhang and Wu 2000).

In the case of the cavern at Sklope dam site (Croatia), left bank, the grout curtain route was re-designed and shifted downstream (Pavlin 1970). Presently the large cavern is filled with water.

One of the first treatments of karst channel flows in the Dinaric karst area was performed on the section of Slano Reservoir grout curtain. This channel flow was plugged at depths of 90–100 m below the surface, about 30 m below the lowest level of water table. The successful injection of the grouting mix could be carried out only during the lowest water table time. Under high water table conditions the groundwater flow was too fast and caused instantaneous washing out of the gravel and grout mix. As a consequence plugging which was initiated in 1966 was carried out intermittently over a period of 4 years and 5 months. Through a borehole Ø 146 mm 4,124 m^3 of gravel and 4,500 tons of grout mix (only dry component) have been injected into the karst channel (Vlahović 1983).

The plugging procedure of karst channels with water flows was successfully used in some other cases: Charmine Dam (France), Krupac Reservoir (Montenegro), Buško

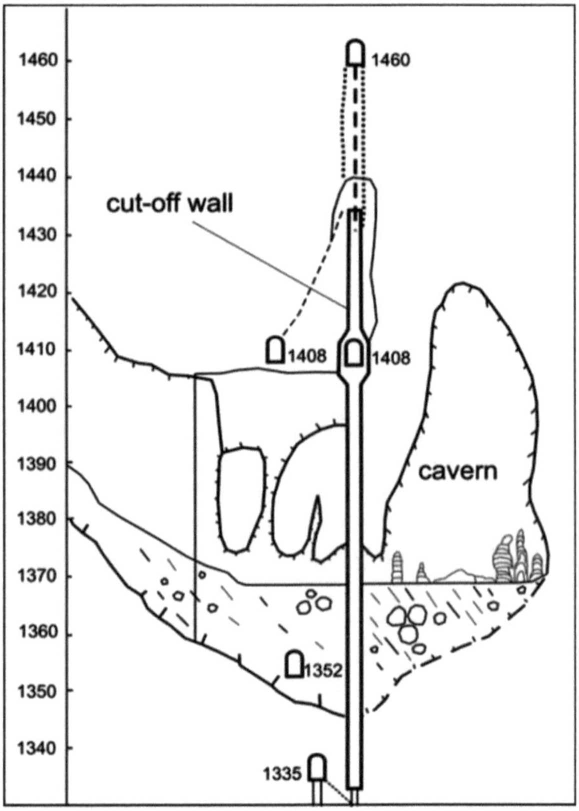

Fig. 17 Wulichong underground dam (China). Concrete cut-off wall (Kang and Zhang 2002)

Blato Reservoir (BiH); Čapljina underground Power Plant (BiH); Guntersville Dam (USA), and Douglas Dam (USA).

Cut-off (Diaphragm) wall is a very effective watertight structure for plugging the highly karstified and wide tectonized zones.

Different technologies are available: deep trenches made by cutters (Gotvand, Iran); overlapping piles (Khao Laem Dam, Thailand; Akkopru Reservoir, Turkey; Pavlovskaya Dam, Russia; Erevan Dam, Armenia; Baipazinskaya Dam, Tajikistan; Wolf Creek Dam, USA); Kamthikhera Dam (India); Khoabin Dam (Vietnam) and a mining method in the form of trenches between the close spaced galleries (Karun I Dam, Iran).

One of the largest cut-offs has been constructed as part of the Wulichong underground dam (Fig. 17). A reinforced concrete cut-off wall is 100.4 m high, 50–30 m wide, and 2.5–2 m thick (Kang and Zhang 2002). For the cut-off foundation, 14,775 m^3 of karstified rock mass and cavern deposits was excavated and replaced with 15,152 m^3 of concrete.

Bath-tub structure means construction of an impervious structure by combination of vertical, inclined or sub-horizontal (Oymopinar Dam, Turkey; El Cajon Dam, Honduras).

Fig. 18 River Trebišnjica, BiH. River bed covered by shotcrete

6.4 Surface Impermeabilization: Structures and Technologies

To reduce water losses from the reservoirs situated in karst the common applied surface structures and technologies are: compacted clayey blankets; different kinds of geomembranes; shotcrete (in the case of exposed rock); plugging of ponors (swallow holes); impermeabilization of ponor zones by grouting blankets; heavy reinforced concrete slabs (in the case of high water pressure); cylindrical dams around the large ponors and estavelles; non-return valves to prevent uplift destruction (in the case of estavelles); construction of rock-filled or earth-filled dikes to amputate sinking zones from reservoir and construction of aeration pipes to prevent air-hammer destruction. In many cases, for a successful water losses reduction strategy, the combination of a few presented approaches is needed.

In the case of Karacaoren II Reservoir (Turkey) three types of protection have been successfully applied: thick concrete, shotcrete and clay blanket (Okay and Soydam Bas 1999).

A few different types of water-tightness protection of the Hutovo Reservoir (BiH) have been applied: compacting of natural alluvial bottom; plugging of individual ponors; grouting of largest ponor zone below the alluvial cover; geomembrane at critical areas; and reinforced shotcrete over the limestone banks.

To prevent leakage along the Trebišnjica lost river (Herzegovina, length of 65 km) 2.2 million m^2 of shotcrete (5 cm thick) have been used (Fig. 18). In natural conditions the river bed was completely dry and the groundwater level was deep below the bed. During the rainy period some sections of river bed are under strong uplift. To prevent destruction of shotcrete the non-return valves were installed.

However, in karst, the risk is never completely eliminated. During reservoir operation any impermeabilization structure is exposed to heavy water pressure and different deterioration processes: piping, erosion, groundwater uplift, and air-hammer and water-hammer effect. Collapses and wide open cracks can occur below the watertight structures (Fig. 19).

To prevent a destructive effect of pressurized air, construction of aeration pipes is necessary (Fig. 20).

Fig. 19 Reservoir bottom (Hutovo, Herzegovina). Collapse and cracks beneath the geomembrane

Fig. 20 Aeration pipe at the reservoir bottom (Hutovo, Herzegovina)

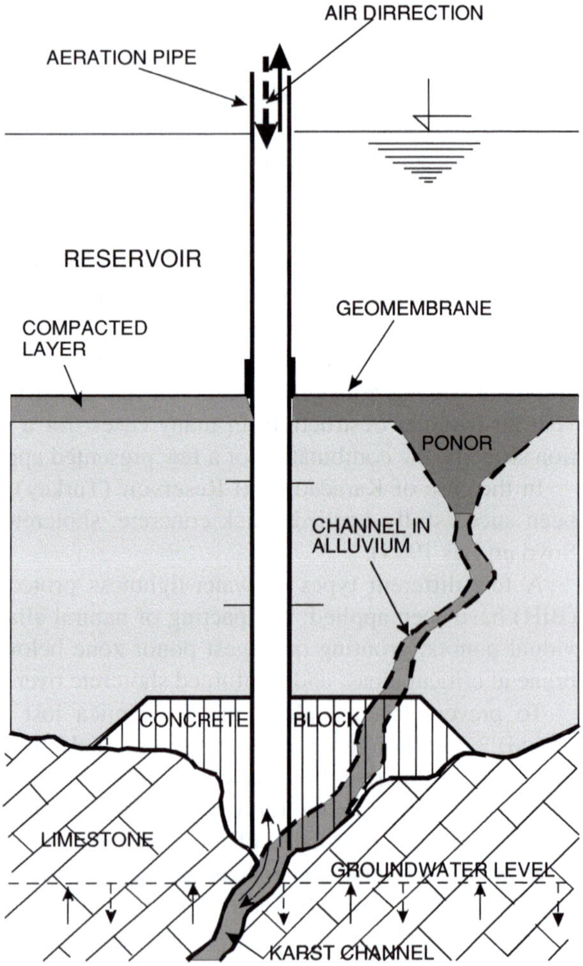

Fig. 21 Nikšić Polje, Montenegro. Cylindrical dam around a large estavelle at the reservoir bank

Fig. 22 China, Guangxi. Gravity concrete dam in front at the entrance of the large underground river

One of the extremely massive structures was constructed to prevent water leakage from the Akkupru Reservoir (Turkey). A 1 km wide stripe of karstified reservoir bottom is protected with one meter of heavily reinforced concrete slab to be resistant against 100 m of water column. The reservoir bank is protected with massive reinforced concrete, also up to the reservoir level of 100 m. To prevent leakage beneath the concrete slab the two rows of cut-off walls (overlapped piles) have been constructed along the reinforced slab, down to the impervious flysch (Günay and Milanović 2005).

To solve the problem of large ponor zones along the perimeter of the reservoir long dikes were used. In some cases to amputate the ponor zone from the reservoir, the dike almost 3 km long has been used (Buško Blato, BiH). In the case of large estavelle or ponor, with a single opening, cylindrical dams were constructed (Fig. 21).

To prevent leakage from the reservoir area in China (Guangxi), the concrete gravity dam was constructed (Fig. 22).

7 Underground Damming

The case studies performed in different regions in the world, particularly in China, provide that an artificial underground storage in karstified rocks may be technically realistic. According to Lu (1986) more than 20 underground reservoirs have been created in different karst regions of China. According to Yuan (1990) in the Xiashi district (Guizhou Province), 16 underground dams have been constructed in karsified rocks. One of largest underground reservoirs was formed on the Linlangdong ground river by construction of a 15 m high dam (Q average is 23.8 m^3/s).

Two underground dams are constructed in the submarine spring karst `channel to mitigate influence of sea water (Port-Miou and Bestouan, France). The first dam is located 2,230 m from the entrance, 147 m below sea level. The other dam (Bestouan) is at a depth of 31 m below sea level and at a distance of 3 km from the channel entrance (Potie et al. 2005).

The project of the Ombla underground dam and reservoir is one of the largest in the world. Location of the underground dam site is proposed about 200 m behind the large Ombla Spring near Dubrovnik (Croatia) at sea level. The average discharge of the spring is $Q = 24.4\ m^3$. The crest of the underground dam is foreseen to be at elevation 100–130 m. Estimated underground operational storage space is about 5 million m^3.

Some dams in China are successfully constructed in front of the large cave to create a reservoir in front of the ponor zone.

8 Problems in Evaporates

More than 60 dams and reservoirs along the world needs rehabilitation because of problems related to the solubility of evaporates, mostly gypsum. A number of cases were reported in the U.S., China, Russia, Iran, Iraq, Argentina, Guatemala, Switzerland, Peru, Venezuela and some other countries (Cooper and Calow 1998).

Some prominent examples are: Mosul Dam (Iraq), Bratsk and Kamskaya (Russia), McMilan Dam (U.S.A.) and Huoshipo Reservoir (China). Grouting in the jointed gypsiferous rock is a questionable and risky task. In many cases, in spite of massive and long-term grouting results, they are not successful (Mosul Dam). According to Maximovich (2006), successful grouting of the gypsiferous foundation rock (Kamskaya Dam) has been done by applying an oxaloaluminosilicate solution.

The salt rocks, which are more soluble than gypsum, are present in the foundation of the Rogunskaya and Nuretskaya dams in Tajikistan. If reservoir water is in touch with salt, the problem of pollution appears as crucial (Gotvand Reservoir, Iran).

The worst was the failure of the St. Francisco Dam (California, US—1929), which killed 450 people. Catastrophic failure of San Juan earth Dam (Spain) occurred during the first filling of the reservoir in 2001. Due to intensive dissolution of gypsum the part of dam collapsed provoking a huge flood in the downstream area (Gutierrez et al. 2003).

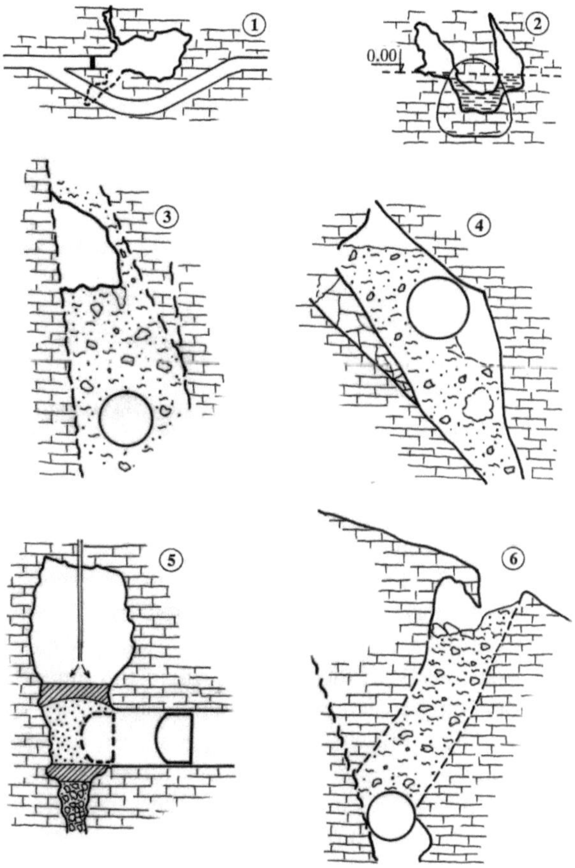

Fig. 23 Different examples of tunnels situated in cavernous rocks

9 Underground Excavations in Karst

Tunnels in karst appear to be the most vulnerable structures. The large caverns and huge inflow of underground water considerably retarded excavation of the derivation tunnels, dewatering tunnels, as well as water transfer and communications tunnels. Some common situations are presented in Fig. 23: (1) excavation of a bypass around a huge cavern; (2) crossing the cavern at sea level; (3) and (4) crossing a cavern filled with unconsolidated deposits; (5) treatment of an empty cavern in front of TBM; (6) collapse at surface provoked by tunnel operation.

From a hydrogeological point of view tunnels are situated: above the maximal groundwater level (GWL); in the zone between minimal and maximal GWL (temporary flooded); and below the zone of minimal GWL (in saturated zone).

Every large sinkhole above the tunnel route is a potential swallow hole during heavy rains. In that case sudden and large water inflow is possible in spite of the fact that the tunnel route is situated high above the GWL.

Particular problems may be encountered during excavation of tunnels below the water table. In this case all karst channels are subject to high pressure, destructive

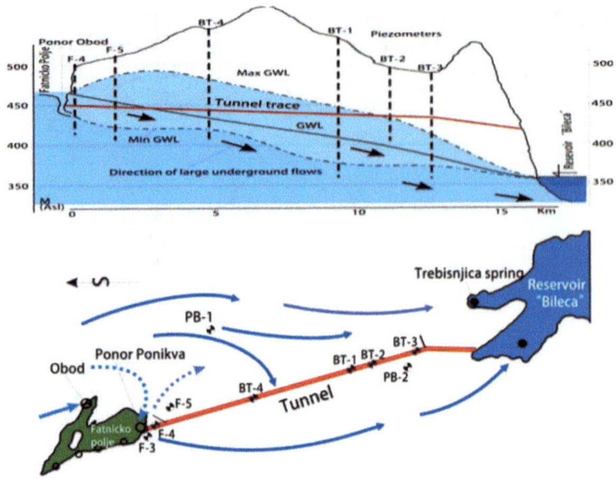

Fig. 24 Tunnel Fatničko Polje—Bileća reservoir

effects of turbulent inflows and an enormous amount of water. Without the possibility of draining the tunnel by gravity, any groundwater intrusion more than 100 l/s is a very serious and dangerous problem. Sometimes consequences are tragic.

Defects during the tunnel operation are very common, particularly in the case of power plant headrace tunnels. These tunnels are under pressure, and failures occurred mostly in the sections where the tunnel intersected large caverns filled with nonconsolidated cave sediments.

The large inflow of underground water considerably retarded excavation of a number of tunnels in different karst regions. Excavation of the Kuhrang III (23 km long tunnel in Iran) had a few years construction delay because of enormous groundwater intrusions (and floods) at a number of sections along the tunnel route situated in limestone. In some sections the pressure of underground flows was 10–11 bars. In the tunnel section with overburden of 1.100 m the karst channel (aperture 1–1.2 m) has been cut by TBM. Discharge of a few liters of muddy water started immediately. After 4 h, discharge increased up to 1 m^3/s. During 24 h, more than 1,000 m^3 of boulders, gravel and sand, including ~500 tons of suspended material in water, was transported from underground into the tunnel. Very complicated drainage structures in front of the TBM head, including a 4 m thick concrete plug, were constructed to allow the plugging of the karst channel, grouting the karstified rock mass in front of the tunnel head and further excavation. In spite of a few horizontal pilot boreholes in front of the tunnel head, the existence of the large karst channel was predicted partially, only.

The intake structure of the water transfer tunnel from the Fatničko Polje to Bileća Reservoir (15.6 km long, BiH) is located at the temporary flooded karst polje at the ponor zone with a swallowing capacity of over 100 m^3/s (Fig. 24). The nine large caverns and a few karstified sections presented a great obstruction for TBM technology. Different unconventional technologies were applied to overcome problems with the cavern. In the dry period, the GWL is below the tunnel level. After heavy rains the groundwater rises above the part of the tunnel in only a few hours. As a consequence of intensively karstified surrounding rock the amount of groundwater intrusion into the tunnel was sudden and large. Beside the direct inflow of sinking water in Fatničko Polje, the large amount from a remote part of the catchment area appears in the middle section of the tunnel.

For optimal excavation planning and to protect people in the tunnel the surrounding catchment area and karst aquifer must be under severe hydrogeological and hydrological monitoring.

According to present experience, TBM technology has a considerable deficiency for application in heavily karstified rocks. Every cavern, empty or filled with clay, with aperture more than 5 m, is a great obstacle for TBM. The problem increases tremendously in the case of a cavity with strong water inflow at the tunnel head. If the cavity is filled with plastic clay, the efficiency of TBM is very low and the possibility for TBM head sinking is very high. In many cases excavation of the by-pass adit around the TBM head for manual cleaning and plugging are the only possibility. This is a time consuming procedure.

To resolve the problem of groundwater burst (6.5 m^3/s) into the 6.17 km long Sozina traffic tunnel (Montenegro) the drainage tunnel (1.75 km long) was excavated below the main tunnel.

Because of the intensive washing process of the clayey/sandy deposits, the tunnel tube of the 8 km long head race tunnel of the Čapljina Reversible Power Plant lost support at a length of 16 m. The leakage of 1 m^3/s was a consequence of lining destruction. Repair works consist of construction of a reinforced arch beneath the tunnel and filling of the empty space by gravel and grout mix. The empty space above the tunnel was left untreated.

Relocation or deviation of the tunnel route because of huge caverns or cavernous zones is common in karst: tunnel for reversible PP Bajina Bašta (Serbia), Čapljina RPP (Herzegovina), Ybing-Gonxian Railroad tunnel, Nunning-Kunming railway tunnel, water intake tunnel in Ontario (U.S.A.) and many others. Due to enormous inflow in the head race tunnel in the case of Seimareh Power Plant the inclination of the tunnel route was changed. In the case of Bajina Bašta head race tunnel the proposed route was changed to avoid the cavernous zone on the base of geophysical investigations (geoelectrical sounding and mapping) from the surface.

Collapses at the surface occurred also as a consequence of tunnel driving througn the caverns filled with clayey material, sand and pieces of rocks including pieces 5–50 m^3: head race tunnel for RPP Čapljina (Herzegovina), and Dodoni tunnel (Greece). In the case of Dodoni, tunnel collapses have occurred at the surface, 100 m above the tunnel (Marinos 2005).

Eleven examples of hydrogeological problems during excavation of tunnels in Chinese karst is presented in the book "Prediction and engineering treatment of water gushing and caverns for tunneling in karst "written by Xingrui 2010. The Long Tan traffic Tunnel (8,693 m) crosses the area of the Tanchunguan underground river. More than 770 m of the tunnel route is situated in the huge cavern filled with clay and gravel deposits (Fig. 25).

Two serious problems during construction were successfully solved: foundation in soft clay deposits and inflow of a huge amount of underground water. In the case

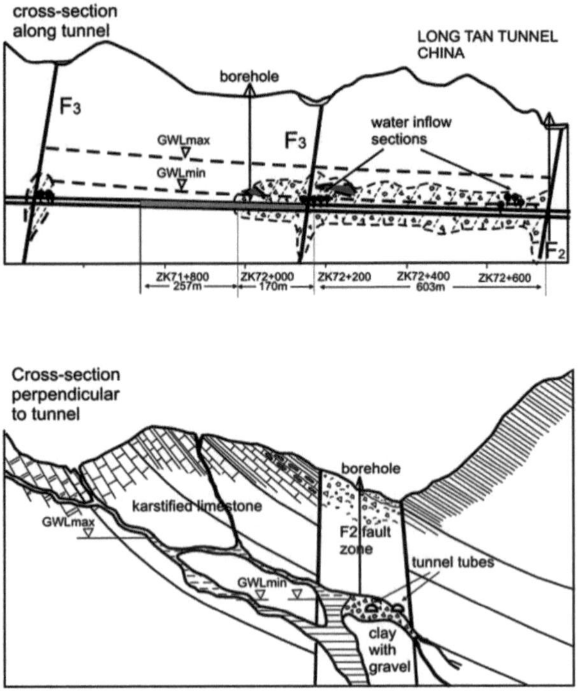

Fig. 25 Ling Tan tunnel near underground river (China) (Xingrui 2010)

of the Zujiayan Tunnel, the problem was the direct connection of the tunnel area with the surface, 200–300 m above, by a few karst channels (shafts). The main water inflow occurred at the tunnel intersection with the karst channel.

10 Induced Seismicity in Karst

In many karst areas local seismic activities are registered during the intensive rainfall when there is abrupt filling of a karst aquifer and rapid rising of groundwater levels. As a consequence of the water table increasing, the pressure of the air in the karst channels and siphons significantly increase. Trapped "air-pillows", including the water in the vapor phase in caverns and fissures, escape the creation of strong explosions that locally can lead to a damaged reservoir bottom and shotcrete lining along the canals. Many times underground air-pillow explosions were recorded by the seismological stations.

One of the earliest documented examples of induced seismicity was registered at 1,837 in an Italian part of Dinaric karst. Explosions of compressed air as a consequence of the fast increasing of water level have been registered by local inhabitants in the Timavo Spring region near the town of Trieste (Galli 1999).

In the other part of the Dinaric karst area (Fatničko Polje, Herzegovina), after heavy precipitation, local inhabitants have noticed strong ground shaking.

Seismic shocks occurred 15–30 h before discharge of the large Obod Spring began (Q_{max} ~ 60 m^3/s). To analyze these events, one seismic station was temporarily installed above the spring (Z-component, 1975). Ground vibrations and shocks, before spring discharge started, were clearly registered by seismograph (Milanović 2000). During abrupt impounding, at the "Bileća" Reservoir (completely situated at extremely karstified rocks) some vibration and shocks were registered by seismograph, which cannot be explained by normal seismic activity or induced seismicity. These vibrations were interpreted as explosions of trapped air and water vapor in karst channels. The sound of escaping air at the surface close to the reservoir and the appearance of colored water (by clayey particles) in the reservoir water were visually observed.

In China, Lu and Duan (1997) distinguished three different types of induced seismicity caused in karstic areas by reservoir water storage: A—loading faulted type (due to reservoir weight); B—pneumatolytic process type (explosions due to uprising of boiling water below the reservoir); and C—cave damaged type (explosions of compressed air mass in caverns).

Similar seismic activity in different karstic regions, related to the fast saturation of a karst aquifer after heavy rainfall, was noticed by authors from China, Italy, Germany and USA.

References

Altug S, Saticioglu Z (2000) Berke arch dam, Turkey: hydrogeology, karstification and treatment of limestone foundation. Present state and future trends of Karst studies. In: Proceedings of the 6th international symposium and field seminar. Marmaris, Turkey

Bianchetti G, Roth P, Vuataz F-D (1992) Deep groundwater circulation in the Alps relation between water infiltration, induced seismicity and thermal springs. The case of Val d'Illieze, Wallis, Switzerland, Eclogae geol. Helv 85(2): 291–305. Birkhauser Verlag, Basel

Bozović A (1985) Foundation treatment for control of seepage, report Q 58. Quinzieme Congres das Grandes Barages, Lausanne

Breznik M (1985) Exploration, design and construction of cut-offs in karstic regions. Quinzieme congres des grandes barrages, Q.58, R.67. Lausanne

Cooper AH, Calow RC (1998) Avoiding gypsum geohazards: guidance for planning and construction. British Geological Survey. Technical report WC/98/5

Djalaly H (1988) Remedial and water tightening Works of Lar Dam. Seizieme Congres das Grandes Barages, San Francisco

Ford D, Williams P (2007) Karst hydrogeology and geomorphology. Willey, NewYork

Galli M (1999) Timavo, Esplorazioni e studi. Suplemento no 23 di Atti e Memorie della Commissione Grotte Eugenio Boegan, Trieste

Günay G, Milanović P (2005) Karst engineering studies at the Akkopru Reservoir area, SW of Turkey. In: International conference water resources and environmental problems in karst, Belgrade and Kotor

Gutierrez F, Desir G, Gutierrez M (2003) Causes of the catastrophic failure of an earth dam built on gypsiferous alluvium and dispersive clays (Altorricon, Huesca Province, NE Spain). Environ Geol 43:842–851

Kang Y, Zhang B (2002) Karst and engineering handling to the karst in Wulichong Reservoir, Yunnan Province. Carsologica Sinica, Institute of Karst Geology, Guilin, China, 21(2): 120–130

Kesić P, Kovačina N (1978) In: Laksiri K, Gunathilake, Iwao Y, Engineering geological evaluation of reservoir leakage problem. 73rd Annual Meeting of ICOLD, 2005 Tehran, Iran, Paper No.: 151–S3
Lu Y (1986) Some problems of subsurface reservoirs constructed in Karst regions of China. Institute of Hydrogeology and Engineering Geology, Beijing
Lu Y, Duan G (1997) Artificially induced hydrogeological effects and their impact of environments on karst of north and south China. In: Proceedings of 30th international geological congress, Beijing, vol. 22, pp 113–120
Marinos P (2005) Experiences in tunneling through katstic rocks. Water resources and environmental problems in Karst. International Conference Belgrade-Kotor
Maximovich NG (2006) Safety of dams on soluble rock (The Kama hydroelectric power station as an example). The Russian Federal Agency for Science and Innovations, Perm
Milanović P (2000) Geological engineering in Karst—Dams, Reservoirs, grouting groundwater protection, water tapping, tunneling. Zebra Publishing Ltd, Belgrade
Milanović P (2006) Karst of Eastern Herzegovina and Dubrovnik Littoral. ASOS, Belgrade in Serbian
Okay G, Soidam Bas A (1999) Experiences on two karstic dam sites. International commission on large dams (ICOLD), Antalia
Pavlin B (1970) Kruščica storage basin in the cavernous Karst Area. Dixieme Congres des Grandes Barages (ICOLD), Montreal
Potie L, Ricour J, Tardieu B (2005) Port-Miou and Bestouan freshwater submarine Springs (Cassis-France) investigations and works (1964–1978). In: Proccedings of water resources and envirnomental problems in Karst. Belgrade-Kotor
Riemer W, Gavard M, Soubrier G, Turfan M (1997) The Seepage at the Atatürk Fill Dam. Dix-Neuvieme Congres des Grands Barrages, Florence
Ruichun X, Fuzhang Y (2004) Karst geology and engineering treatment in the Geheyan project on the Quingjiang River, China. Engineering Geology 76:155–164, Elsevier
Trzhtsinsky Yu B, Filipov VM (1981) Tectogenic activization of Karst in Angara Water Reservoir. International engineering geology symposium, Istanbul
Vlahovic V (1983) Karst Reservoir of Slano. The Montenegrin Academy of Sciences and Arts. Special editions, vol. 14. Titograd (Podgorica)
Xingrui H (2010) Prediction and engineering treatment of water gushing and caves for tunneling in Karst. Guangxi Normal University Press, China
Yuan Daoxian (1990) Construction of underground dams on subterranean streams in South China Karst. Institute of Karst Geology, Guilin
Zhang B, Wu M (2000) Seepage control treatment of Blind Valley Reservoir, Gao Yaoji, China. Hydroelectric Power Construction Engineering Consultant Company, Yunnan Province

Inferred Conduit Network Geometry from Geological Evidences and Water-Head in a Fluvio-Karstic System (Val D'Orleans, France)

Chrystelle Auterives, Stéphane Binet and Patrick Albéric

Abstract Flow modeling is a useful tool to investigate sustainable scenarios of water use for groundwater resources managers. However, the applications of this approach in karstic aquifers are still limited due to the difficulties of locating and describing the position and geometry of conduits. In the karstic aquifer of the 'Val d'Orléans' (France), the location of the conduits was identified by the presence of land-surface collapses and voids in the boreholes. In this dataset, 103 logs showed voids >0.1 m and a highly porous zone around 80 m.a.s.l. In this chapter, a methodology is proposed to interpolate the dataset to build a karstic network at the regional scale. Here, four scenarios of the conduit network geometry are tested, related to the morphology of the network and its complexity. The scenarios are included in a 3D coupled continuum-conduit flow model (Feflow®). The flows in the conduits and in the host rock are respectively described by the Manning-Strickler equation and Darcy's law. Constant boundary conditions and hydrodynamic properties are assumed to test the model sensitivity to different network geometry scenarios. The most relevant scenario is selected by comparing calculated and observed water-heads in the boreholes. It aims to calculate a water balance at the regional scale, results suggest that accurate models (correlation coefficient $r^2 > 0.9$) can be obtained with an average hydraulic diameter approach and with highly simplified conduit network geometries. This confirms the interest of discrete continuum approaches, even if the actual conduit geometry will never be known.

Keywords Fluvio-karstic system • Karstic network • Conduit flow models • Conduit network geometry • Val D'Orleans • France

C. Auterives · S. Binet (✉) · P. Albéric
Institut des Sciences de la Terre d'Orléans—UMR 7327, Université d'Orléans, CNRS/INSU, 1A, rue de la Férollerie, 45071 Orléans, France
e-mail: stephane.binet@univ-orleans.fr

J. Mudry et al. (eds.), *H2Karst Research in Limestone Hydrogeology*, Environmental Earth Sciences, DOI: 10.1007/978-3-319-06139-9_3,

1 Introduction

To model the groundwater flows in a karstic aquifer, the modeler is confronted with the duality of flows, with the major role of the permeable discontinuities as the water transfer compartment and the important role of the surrounding rock as the water storage compartment (Kiraly et al. 1995). This duality is enhanced by the existence of turbulent flows (Cheng and Chen 2004) that may take place in the conduits. To integrate this duality, numerical schemes of coupled continuum-conduit flow models were developed. However, the accuracy of the groundwater calculations is limited by the difficulties of describing the location and the geometry of the conduits.

Some global conceptual models were proposed to estimate the average properties of the conduit at the regional scale from discharge measurements and recession coefficient (Fornillo 2011), or from water quality (Grasso et al. 2003). Other studies estimate the water exchanged between the discontinuities and the surrounding rock from hydrogramm and chemiogramm (Binet et al. 2007; Charmoille et al. 2009).

In aims to proposed distributed models, several solutions have been developed to extrapolate network geometry from geostatistic and stochastic models (Fournillon et al. 2010; Borghi et al. 2010) or from geomorphological interpretations and field evidences (Perrin and Luetscher 2008; Binet et al. 2010). In these studies, the accuracy of the network description is related to the available dataset. The actual geometry will never be modeled, and it is difficult to evaluate which accuracy of the network geometry is needed for proposal of a groundwater model calibrated at the regional scale.

This chapter aims to use 2,000 boreholes and 24 artificial tracer tests from a well-known conduit flow system (the Val d'Orléans karst system) to infer a conduit network geometry and to test the sensitivity of the karstic network geometry at the regional scale on the groundwater model results.

2 Hydrogeological Setting

The Val d'Orléans is considered as a vast depression of the major bed of the Loire River, 37 km long and from 4 to 7 km wide (Fig. 1). The karst aquifer is hosted within a carbonate lacustrine deposit called the Beauce limestone overlain by the Quaternary alluvia of the Loire River. In some places, a clay layer is interbedded, creating a confined area in the limestones. This geological setting creates a multilayeed aquifer system with no neglecting flux between the alluvia and the limestones (Lepiller 2006).

The Loire River feeds more than 80 % of the water hosted in the carbonated karstic aquifer (from 11.5 m^3/s during low water period). Rainfall recharge is 150 mm/year (Gutierrez and Binet 2010). The water flows from the city of Jargeau towards several springs of the Loiret River through the under pressure karst

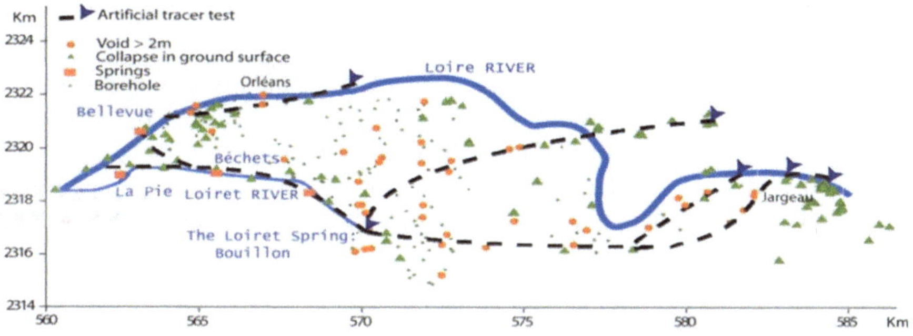

Fig. 1 Map of the Val d'Orléans Aquifer. Observed voids in the boreholes (*circles*) and springs (*squares*), land-surface collapses (*triangles*) and the tracer tests (*dashed lines*) are indicated

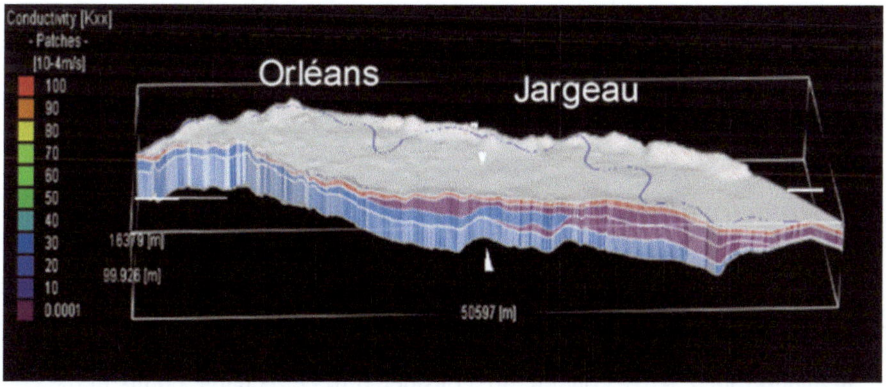

Fig. 2 3D model of hydraulic conductivities of the surrounding rock in the Orleans aquifer. Alluvia deposits correspond to the red layer, the limestones and clays are respectively indicated in blue and purple. Vertical exaggeration: X 100

networks (Fig. 1; Lepiller 2006; Alberic and Lepiller 1998). The Loiret springs are considered as the main emergence of the water infiltrated from the Loire River (from 0.1 to 5 m^3/s). In this area the karstic conduits, explored by speleologist divers, present a global flow rate of 2–10 m^3/s.

Piezometric maps of the study area (based on more than 700 boreholes) have been established for periods of low and high water levels (Desprez 1967). The annual difference between these two levels is 1 m.

Martin and Noyer (2003) synthesized the hydrogeological properties of the geological layers. Pumping tests in the limestones show a hydraulic conductivity of c. 5–30 $\times 10^{-4}$ m/s. The alluvia deposits present a hydraulic conductivity ranging between 10–120 $\times 10^{-4}$ m/s. The hydraulic conductivity of the interbedded clay deposit is assumed to be 10^{-8} m/s. The 709 wells used for irrigation and water supplies were counted (Fig. 2). The total amount of pumped water is about 3×10^7 m^3/year (Binet et al. 2010).

3 Materials and Methods

The 3D geological model and the location of the conduits is interpreted from two databases (borehole logs and land-surface collapses realized by BRGM, the French geological survey (http://infoterre.brgm.fr/ and http://www.bdcavite.net/). In the Val d'Orléans aquifer, 2,000 lithological borehole logs were analyzed to extract information about geology and underground karstic voids.

3.1 Geological Model

The thickness of each geological layer (alluvia, clays and limestones) observed in the boreholes is spatialized using a kriging method and a variogram analysis, with the GDM software (www.gdm.brgm.fr). The interpolated thicknesses, adjusted from the variogram, are cross-validated. If two boreholes are located in a 50 m neighborhood borehole, they are merged. The result of the geological model is scaled for a 250 m × 250-m square mesh.

3.2 Geometry of the Karstic Network Model

The voids recorded in the borehole logs are considered to belong to the karstic conduit. The position (X, Y) of the voids >2 m are presented in Fig. 1. The 147 land-surface collapses (Fig. 1, from the cavity database), related to the presence of an underground active conduit, are considered to be recharge points in the conduit network. Based on 24 artificial tracer tests (Lepiller 2006; Joodi et al. 2009), observed voids have been shown to be connected as a part of the karstic conduit network (Fig. 1). The tracer tests evidence the output points of the network.

To create a model of the conduit network geometry, a triangle interpolation between the observed voids was used to link the voids between them and a mesh used for design the network geometry was created (Fig. 2). The conduit network geometry is considered to link the recharge and the output points within this mesh. Numerous plausible network scenarios can be proposed. Four scenarios of the network geometry were tested in this chapter with the aim to test the sensitivity of a regional groundwater flow model to the network geometry.

3.3 Groundwater Model

For the second step, the four networks are imported in a coupled continuum-conduit flow model (Felflow®). The 3D groundwater model includes discrete conduits with turbulent flows described with a Manning-Strickler law hosted in a surrounding rock where the flows are described with a Darcy Law. In Orléans, all the limestone thickness is always under saturated conditions.

Constant heads are fixed in the Loire and in the Loiret rivers. The hydraulic head boundary conditions of the limestone layer are assumed constant to take into account the fact that the water from the Beauce and Sologne area (lateral flows) feed the Loiret springs. The roughness coefficient is adjusted to fit the maximum flow velocities observed from the artificial dye tracer tests. Hydraulic conductivities of the rock are assumed homogeneous in a layer. The water mass balance is adjusted changing the average conduit diameters.

The land-surface collapses, considered as vertical conduit, have the same properties than the horizontal conduits.

The accuracy of the network geometry scenarios is tested by comparing measured and calculated water tables. The field data includes 300 water-heads measurements performed in September 1966 (Desprez 1967). Those measurements are compared with a steady-state calculation realized for each of the scenarios. The correlation coefficient R^2 between the measured and calculated water table to assess the accuracy of the scenarios was used.

4 Results

4.1 Geological Model

The hydrogeological model presents three lithology types and 5 layers (Fig. 2). The total thickness can reach 50 m. The position of voids is restricted to the eastern part of the area, where the limestone layer is not covered by clay layer. When the clay layer disappears, the limestone becomes connected with the surface water leading to the dissolution of the limestone and the development of conduits.

4.2 Geometry of the Karstic Network Model

4.2.1 Elevation of Conduits

Out of 2,000 borehole logs, 103 logs present voids >0.1 m (Fig. 3). The field data show that the mean elevation of the voids is c. 80 m. a.s.l (c. 10–20 m below the ground surface). The voids have a mean diameter of 3.5 m. A corresponding cross-section of about 10 m^2 is estimated by assuming a circular shape of the conduit (Fig. 4) which is located in a zone around 70–80 m a.s.l.

4.2.2 Geometry of the Conduit Network

Here, four network scenarios are used from a simple one to a more complex one that honors all the dataset of karst evidences known in the Val d'Orléans. The first model includes a single conduit (Scenario 0) between the Loire River

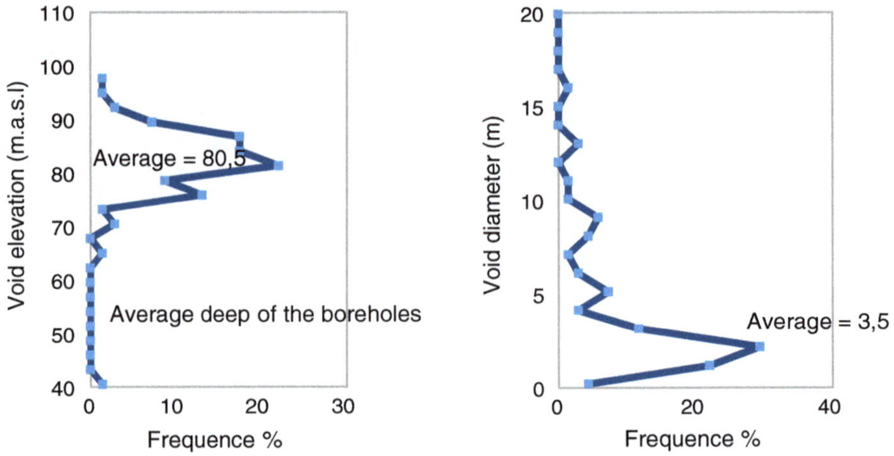

Fig. 3 Frequency repartition of voids observed in the boreholes related to a/the elevation, b/their diameters

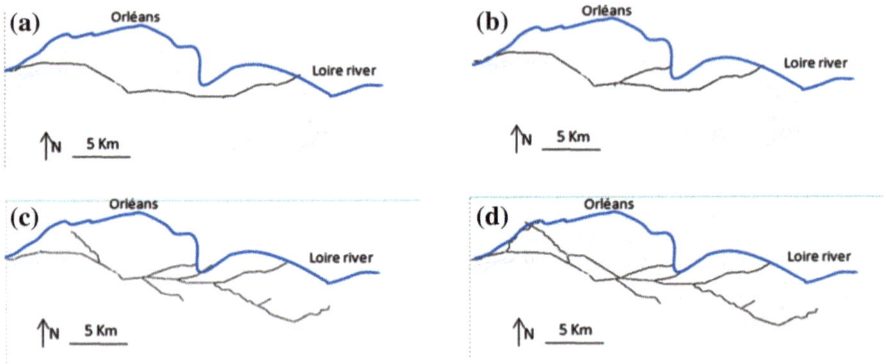

Fig. 4 Presentation of the four network geometry scenarios tested to describe the groundwater flows in the Val d'Orleans. **a** Single conduit. **b** Network deduced only with major voids. **c** Network with secondary tributaries. **d** Network with an anastomosed geometry

and the Loiret River. Scenario 1 is a simple network to link all the voids >2 m. Then a dentritic geometry network is proposed to link the voids <2 m to the voids >2 m (Scenario 2). The last Scenario shows an anastomosed geometry network (Scenario 3). Scenarios 1 and 2 h both the voids and tracer test data, and Scenario 4 also includes the land-surface collapses data. For the four scenarios, an average constant hydraulic area is assumed for the overall conduits.

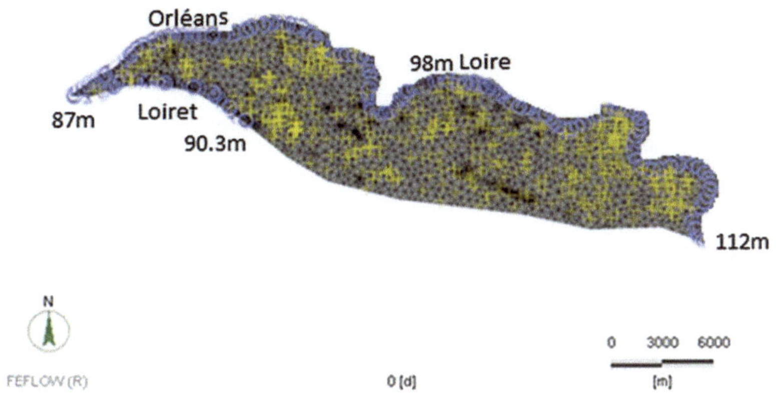

Fig. 5 Boundary conditions in the alluvia layer 1 and mesh of the Val d'Orléans model. *Circles* represent constant head, *black triangles* are mesh; rainfall recharge (source term) is 150 mm/year; cross are pumping wells

4.3 Groundwater Model

Figure 5 shows the first layer with the boundary conditions applied in the flow model. Each layer is described with mean values of hydraulic conductivities proposed by Martin and Noyer (2003). The constant heads, deduced from measurements data (French Environmental Agency), range between 87 m down to Orléans to 110 m (Fig. 5). Location of the pumping wells is represented by crosses. The mesh used for the groundwater model is represented by the black triangles.

4.3.1 Flow Velocity in the Conduits

The flow velocities calculated in the conduit are a key parameter of discrete-continuum model. To adjust the roughness coefficient, calculated maximum flow velocities are compared with the maximum flow velocities obtained by artificial tracer tests (Joodi et al. 2009). The field data from tracer tests (Fig. 6) validate that the Manning-Strickler law is adapted for the flows in this kind of fully saturated karstic conduits. They show that the field data are coherent with a 15 $m^{1/3}$ s^{-1} roughness coefficient.

4.3.2 Sensitivity Analysis

Table 1 shows the accuracy of each geometry network scenario (proposed in Fig. 3) tested by calculating the R^2 of the measured/calculated water-head in the aquifers. To respect the water balance described in Sect. 2, the average area of the conduit is adjusted. The adjusted values are shown in Table 1.

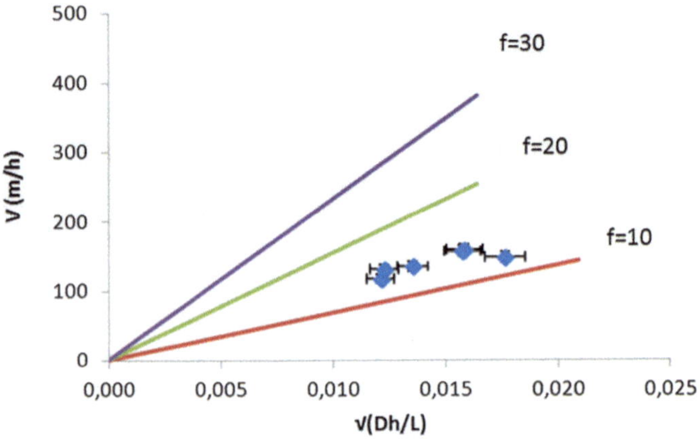

Fig. 6 Maximum flow velocities in the conduits versus the roots of the hydraulic gradient (according to Manning Strickler Law). Observed data from artificial tracer tests versus calculated results for a 5, 10 $m^{1/3}\ s^{-1}$ roughness coefficient

Table 1 Correlation coefficient R^2 for the four scenarios of network geometry, calculated water balance and average diameter of the conduits chosen for the model. (All the other parameters are fixed as described in 3.3)

Scenario	0	1	2	3
Water table R^2	0.6	0.65	0.8	0.97
Water balance (m^3/day)	1.2×10^6	1.2×10^6	1.2×10^6	1.2×10^6
Average area of conduits (m^2)	45	20	15	10

Scenario 3 is identified as the best model which respects all the field data (Table 1). Scenarios 2 and 3 present a high R^2 coefficient, and can be therefore considered more closely to the actual geometry of the conduit. The existence of an anastomosed geometry (Scenario 3) is confirmed, although the R^2 does not increase strongly, this is the only solution that honors average area of the conduits observed in Sect. 4.2. Under fixed hydrogeological properties and boundary conditions, the model sensitivity (R^2) varies from 0.6 to 0.95 only by changing the geometry network. This shows the sensitivity of the flow model to the network geometry.

Note that a simple conduit network (Scenario 0) enables to reproduce the water balance and gives a 0.6 r^2 value for water table simulation. Related to the database available, a simple conduit can produce a regional groundwater flow model with a good accuracy.

The observed versus calculated water heads for the 300 boreholes are presented in the Fig. 7. This scatter reports to Scenario 3, that is the best fit obtained. The observed water heads are reproduced with an error range about ±50 cm.

This groundwater model was realized in an area with a high density of water head measurements (>1 boreholes/km^2) and shows that a simple average area for

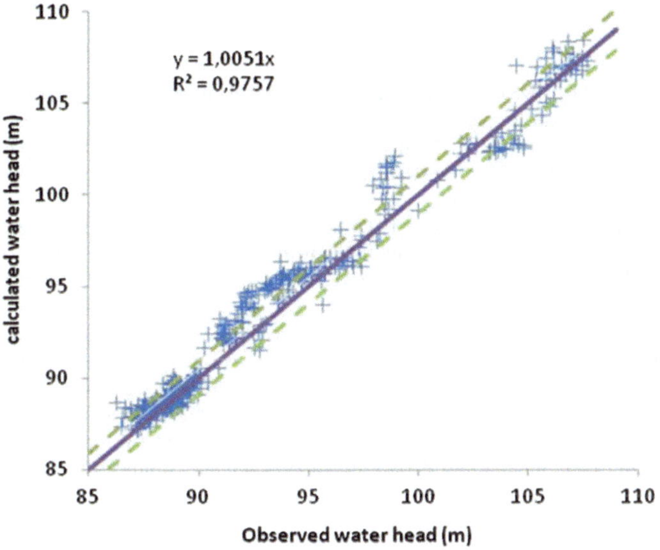

Fig. 7 Observed versus calculated water head in the Val d'Orléans aquifers with the scenario 3. 1:1 line and plus or less 0.5 m lines are plotted

the conduits and few hypotheses about network geometry gives accurate model ($r^2 > 0.8$) at the regional scale. Even if the real geometry of the network will never be known, a model of karst aquifer at the regional scale based on an averaged conduit area approach and simplifying the network geometry honors in an acceptable way the discharge and water table dataset.

5 Conclusion

The conduits in the karstic aquifer of the 'Val d'Orléans' (France) are located in a highly porous zone around 80 m.a.s.l. characterized by voids with a mean diameter of 3.5 m. The 3D groundwater flow model is used to select the most realistic geometry of the conduit network from scenarios derived from geomorphological observations. These results show that the anastomosed geometry network is the best scenario. The model is sensitive to changes in the network geometry. In aims to calculate a water balance at the regional scale or to describe a water balance, it appears to be possible to produce accurate models with highly simplified conduit network geometries. This confirmed the interest in the discrete continuum approach, even if the real conduit geometry will never be known.

Acknowledgments This work was founded by the "Région Centre" and the "Département du Loiret" through the TRAC and ICERE projects.

References

Alberic P, Lepiller M (1998) Oxydation de la matière organique dans un système hydrologique karstique alimenté par les pertes fluviales (Loiret, France). Water Resour 32:2051–2064

Binet S, Guglielmi Y, Bertrand C, Mudry J (2007) Unstable rock slope hydrogeology: insights from the large-scale study of western Argentera-Mercantour hillslopes (South-East France). Bull Soc Geol Fr 178:159–168

Binet S, Joodi A, Joigneaux E, Alberic P, Gutierrez A (2010) Localization of reactive transport zone in a saturated karstic conduit highlights from natural and artificial tracer tests. Advances in research in Karst media. Malaga (Spain). In: Andreo B, Carrasco F, Duran JJ, LaMoreaux JW (eds). Springer, Berlin

Borghi A, Renard P, Jenni S (2010) How to model realistic 3D karst reservoirs using a pseudo-genetic methodology-example of two case studies. Advances in research in Karst media. Malaga (Spain). In: Andreo B, Carrasco F, Duran JJ, LaMoreaux JW (eds). Springer, Berlin

Charmoille A, Binet S, Bertrand C, Guglielmi Y, Mudry J (2009) Hydraulic interactions between fractures and bedding planes in a carbonate aquifer studied by means of experimentally induced water-table fluctuations (Coaraze experimental site, southeastern France). Hydrogeol J 17:1607–1616

Cheng JM, Chen CX (2004) An integrated linear/non-linear flow model for the conduit-fissure-pore media in the karst triple void aquifer system. Environ Geol 47:163–174

Desprez N (1967) Inventaire et étude hydrogéologique du Loiret. Report DSGR 67 A21, Ed. BRGM. 125 p

Fiorillo F (2011) Tank-reservoir drainage as a simulation of the recession limb of karst spring hydrographs. Hydrogeol J 19:1009–1019

Fournillon A, Viseur S, Artib B, Borgomano J (2010) Insights of 3D geological modelling in distributed hydrogeological models of kartic carbonate aquifers. Advances in research in Karst media. Malaga (Spain). In: Andreo B, Carrasco F, Duran JJ, LaMoreaux JW (eds). Springer, Berlin

Grasso DA, Jeannin P-Y, Zwahlen F (2003) A deterministic approach to the coupled analysis of karst springs 'hydrographs and chemographs. J Hydrol 271:65–76

Gutierrez A, Binet S (2010) La Loire souterraine, circulations karstiques dans le Val d'Orléans. Géosciences. Ed. BRGM

Joodi A, Sizaret S, Binet S, Bruand A, Alberic P (2009) Development of a Darcy-Brinkman model to simulate water flow and tracer transport in a heterogeneous karstic aquifer (Val d'Orléans, France). Hydrogeol J 18:295–309

Kiraly L, Perrochet P, Rossier Y (1995) Effect of the epikarst on the hydrograph of karst springs: a numerical approach. Bulletin du Centre d'Hydrogéologie 14:199–220

Lepiller M (2006) Hydrogéologie du Val d'Orléans. In les aquifères de France. ed. BRGM—AIH 2006, 956 p

Martin JC, Noyer NL (2003) Caractérisation du risque d'innondation par remontée de nappe sur le val d'Orléans. RP-52121-FR. ed. BRGM. 170 p

Perrin J, Luetscher M (2008) Inference of the structure of karst conduits using quantitative tracer tests and geological information: example of the Swiss Jura. Hydrogeol J 16:951–967

Contribution of Isotopes of the Water Molecule to Determine Recharge Altitude of the Main Springs Welling Up in the Middle Atlas Limestone (Morocco)

Abdelhadi El Ouali, Soumaya Sefrioui, Jacques Mudry, Omar Fassi Fihri, Ali Essahlaoui and Hamid Marah

Abstract The present study contributes to characterize the Middle Atlas limestone karst aquifers by implementing water stable isotope techniques, to determine the recharge altitudes of major springs issuing from the Liassic dolomitic limestone, and helping to delineate protection zones of the springs. Chemical analyses of spring water show a composition that reflects both reservoir lithology and chemical evolution over the flow transit through the reservoir. Analysis of the evolution of deuterium as a function of oxygen-18 suggests a local water line similar to that derived for rainwater at the Fez weather station, indicating that the concerned aquifers are recharged through rapid infiltration, without appreciable evaporation of rainwater. Application of the established −0.27 ‰ per 100 m altimetry gradient for oxygen-18 to the isotopic signatures of other analyzed springs enables the backtracking of recharge altitudes of these springs. Comparison of these results with local and regional hydrogeological data shows a notable agreement among springs for which the drainage basin is known, and allows for more precise localization of recharge area of springs for which the drainage basin is poorly known, with recharge altitudes up to 1,090 m higher than their emergence zones. To a certain extent, these results help to determine recharge zones for Middle Atlas springs and thus help in strategizing protection for this resource.

A. El Ouali (✉) · S. Sefrioui · A. Essahlaoui
Research Team, Water Science and Environmental Engineering, Department of Geology, Faculty of Sciences, Moulay Ismail University, Meknes, Morocco
e-mail: abdelhadielouali@yahoo.fr

J. Mudry
Faculty of Sciences Techniques, Chrono-Environment Laboratory UMR 6249, 16, Gray Road, 25030 Besançon Cedex, France
e-mail: jacques.mudry@univ-fcomte.fr

O. Fassi Fihri
Water Service in Meknes, Secretary of State in charge of Water and Environment, Meknes, Morocco

H. Marah
Laboratory of Isotope Hydrology, CNESTEN, Meknes, Morocco

J. Mudry et al. (eds.), *H2Karst Research in Limestone Hydrogeology*,
Environmental Earth Sciences, DOI: 10.1007/978-3-319-06139-9_4,

Keywords Middle Atlas • Springs • Karst • Chemical analysis • Oxygen 18 • Deuterium • Recharge altitude • Protection zone • Isotopic gradient

1 Introduction

The Meknes-Fez area, which is situated north of the Middle Atlas, is one of the most important agricultural regions in Morocco. Its agricultural and drinking water resources are provided by the Middle Atlas karst massif (Fig. 1), and shallow and also deep aquifers of the Meknes-Fez Basin (Chamayou et al. 1975; Essahlaoui 2000). The average climate of the area is Mediterranean, with Oceanic influences. The mean annual rainfall is 600 mm. It decreases with continentality effect, and increases according to altitude and exposure to ocean disturbances from the west. The highest temperatures can be recorded in July and August, and the minimum ones in January. Yearly average values vary widely between 10 and 20 °C, according to continentality and altitude. The Middle Atlas aquifers are composed of Liassic carbonates, characterized by a permeability hosted in fissures and karst conduits (groundwater of limestone plateaus called 'causses' and folded Middle Atlas (du Dresnay and Studer 1975).

The Liassic calcareous and dolomitic massif gives birth to many springs that emerge at the Northern and Southern of Middle Atlas (North of ElHajeb city) thrust faults. Springs at the focus of the present study are chosen according to their multiple uses, to the magnitude of their flow rates, and to their hydrogeological context, as major outlets draining groundwater. The most important among those are the Ribaa and Bitit springs (average discharge: 1,324 L/s), in the foothills of the Middle Atlas limestone plateau, that supplies Meknes city with drinking water as well as agricultural water.

For many years, the management of these springs concerns the Sebou Watershed Agency and the National Office of Drinking Water (ONEP). Indeed, karsts systems represent a valuable alternative resource to other aquifers over exploitation, and to surface waters, especially in semi-arid Mediterranean environments. Thus, the goal of this study is to contribute in determining the origin of the water outflowing from these high discharge springs, in order to aide in determining their protection zones.

2 Materials and Methods

A sampling campaign of 17 springs was performed during a high water period, in February 2007. Measurements of physical and chemical parameters (temperature, Electrical Conductivity, pH and alkalinity) were performed in situ.

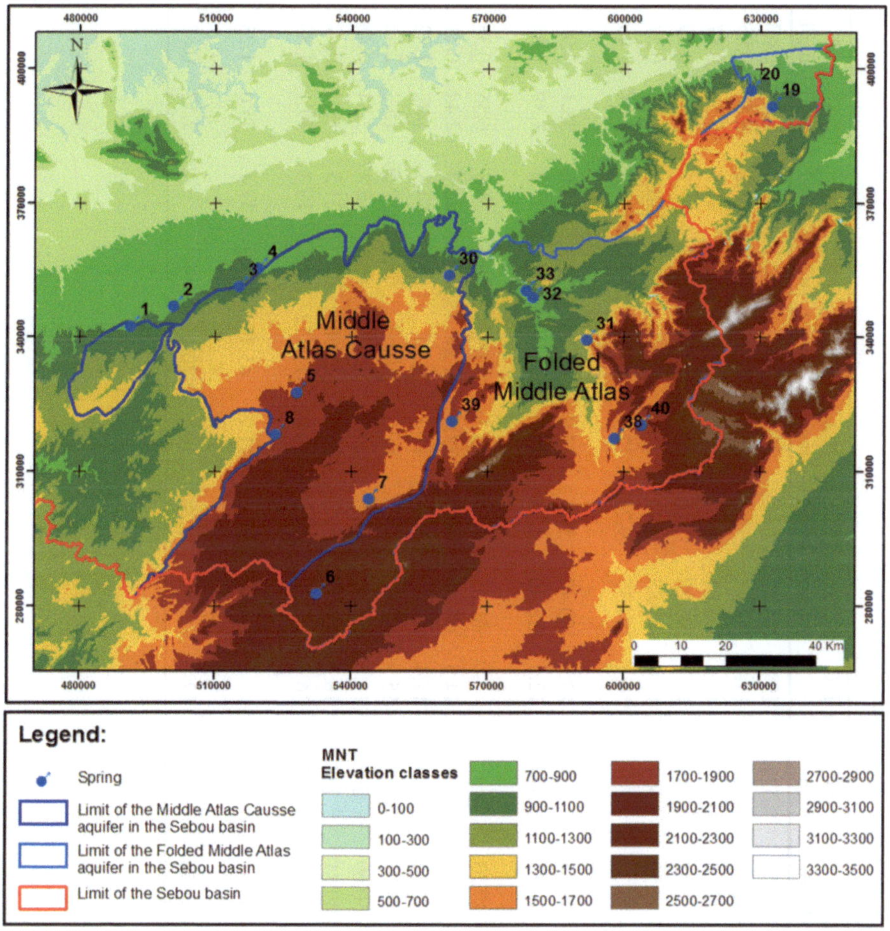

Fig. 1 Location of the study area and spatial distribution of the Middle Atlas groundwaters (Morocco)

Analyzes (Table 1) concerned major chemical elements and stable isotopes of the water molecule (^{2}H and ^{18}O). These analyzes were performed in the laboratory of isotope hydrology of the National Center of Nuclear Study of Morocco (CNESTEN), using a mass spectrometer (Table 1).

The methodology enables the exploration of the potential of multidisciplinary approaches, focused on the use of stable isotopes of the water molecule, to determine the recharge areas of the main studied springs.

Table 1 Water analyses of the sampled springs

N°	Spring	Aquifer	$\delta^{18}O$ (vs. SMOW ‰)	$\delta^{2}H$ (vs. SMOW ‰)	EC μs/cm	T°C	$HCO3^-$ mg/L	Cl^- mg/L	NO_3^- mg/L	SO_4^{2-} mg/L	Ca^{2+} mg/L	Mg^{2+} mg/L	Na^+ mg/L	K^+ mg/L
1	Ain Maarouf	Guigo 'causse'	−6.16	−37.51	537	19.8	365.39	53.5	26.3	8.11	60.1	40.5	32.2	0.67
2	Ain Aghbal	ElHajeb 'causse'	−6.66	−38.66	685	17.7	354.41	13.0	58.9	12.60	69.3	38.2	8.50	0.85
3	Ain Ribaa	ElHajeb 'causse'	−7.42	−42.68	583	16.6	378.81	10.7	11.6	2.56	71.1	31.7	6.56	0.47
4	Ain Bittit	ElHajeb 'causse'	−7.65	−42.87	588	16.7	395.28	10.0	10.3	2.54	62.3	31.7	6.33	0.49
5	Ain Zerrouka	Ifrane 'causse'	−8.03	−47.50	604	12.6	417.85	4.46	7.48	3.23	73.9	30.8	2.39	0.37
6	Aghbalou Aberchane	Guigo 'causse'	−8.90	−55.31	407	10.1	204.35	9.24	12.1	27.58	53.6	16.9	7.44	1.49
7	Ain Titzil	Folded M.A	−8.04	−45.84	669	13.7	199.47	56.5	13.9	69.20	59.2	24.6	38.1	4.96
8	Ain Sidi Rached	Ifrane 'causse'	−8.04	−48.13	577	12.1	418.46	2.84	4.00	1.90	66.3	33.2	1.39	0.33
19	El Anssar	Folded M.A	−7.55	−47.23	–	–	309.88	10.6	15.1	8.09	55.5	16.8	6.55	0.35
20	Ain Ras El Ma	Folded M.A	−7.57	−46.00	499	14.5	387.35	3.02	7.06	7.87	61.1	26.7	2.42	0.41
30	Ain Regraga	Sefrou 'causse'	−7.48	−51.64	1198	17.4	457.5	167.	11.1	9.20	85.6	44.4	146.	1.42
31	Ain Bouk	Folded M.A	−8.78	−55.72	427	15	298.9	3.05	8.35	10.67	69.2	9.66	15.3	2.46
32	Ain Sebou	Folded M.A	−8.32	−50.69	617	15.5	451.4	13.5	4.62	10.79	80.5	8.87	2.74	0.43
33	Timedrine	Folded M.A	−8.33	−49.81	738	18.3	460.55	50.4	3.62	12.58	122	21.5	13.4	0.47
38	Skhounatemar-mocha	Folded M.A	−8.61	−54.87	812	36	273.28	79.2	10.0	135.12	115.	19.3	39.1	1.89
39	SkhounateS. Mdez	Folded M.A	−8.70	−55.69	847	–	376.98	101.	11.7	45.77	77.9	37.4	75.0	1.25
40	Ain Tataw	Folded M.A	−9.53	−60.50	336	12.8	213.5	2.28	10.2	20.16	64.9	9.40	2.17	0.82

causse limestone plateau; *MA* Middle Atlas)

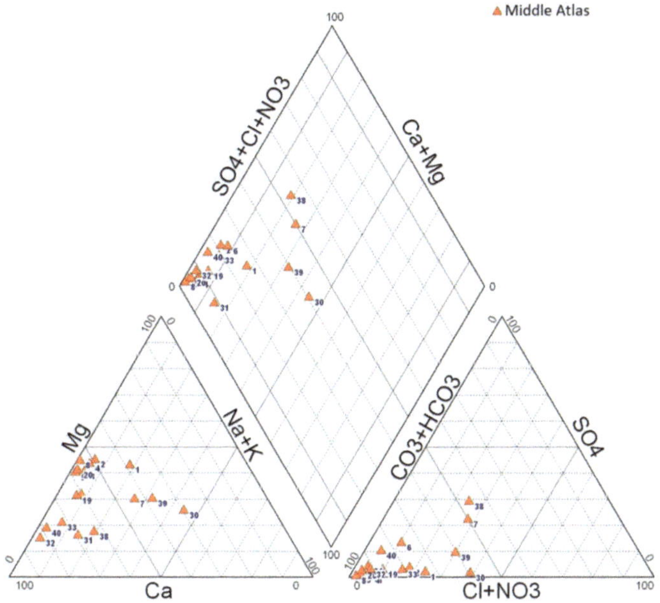

Fig. 2 Piper's diagram of the Middle Atlas groundwaters

3 Results of Chemical Characterization

The results of chemical analyzes of sampled springs are plotted in the Piper's diagram (Fig. 2). This representation allows highlighting a wide variety of chemical types of springs, with a strong prevalence of water with magnesium and calcium bicarbonate types. In the folded Middle Atlas, there is a distinction between springs evolving from two distinct poles. The first one is calcium-magnesium type, in springs emerging from the Lower Liassic dolomitic formations (sample No. 20). The second one is represented by calcium springs emerging from the Middle Liassic limestone (sample Nos. 32 and 40). These two basic types evolve into a sodium pole (sample No. 39) for the first group, and into a sulfate pole (sample No. 38) for the second one. On the Plateaus, there is a noteworthy evolution from springs situated close to their recharge area (sample Nos. 5 and 8) to the more remote springs (sample Nos. 1, 7 and 30).

At the foothills, a significant evolution from the piedmont springs of the 'Causses' (sample Nos. 2, 3 and 4) can be displayed.

3.1 *Relationship* $\delta^2H = f(\delta^{18}O)$

Plotting isotopic measurements (Fig. 3) for 43 different precipitation episodes heavier than 5 mm and spread over the 1994–2002 period on the diagram $\delta^2H = f(\delta^{18}O)$, shows that rainwater fits a Local Meteoric water Line (LML) for which

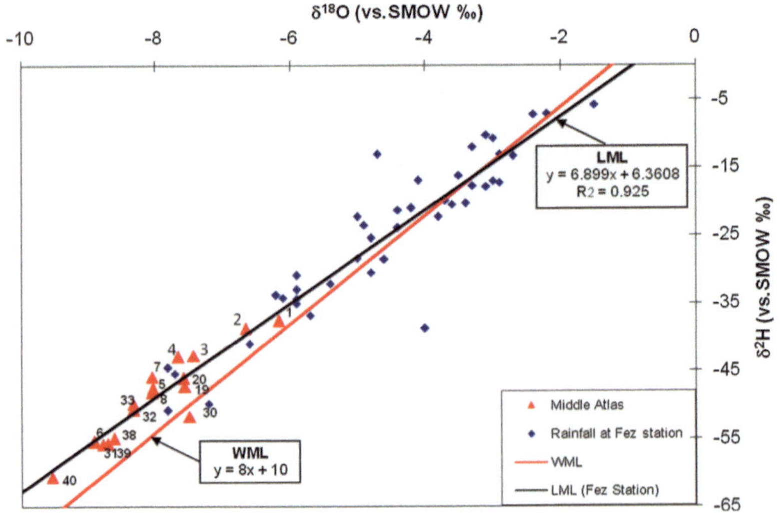

Fig. 3 Variation of ^{2}H content based on ^{18}O in rainwater of the Fez weather station (1994–2002) and in the Middle Atlas spring waters

the equation is: $\delta^2H = 6.9 * \delta^{18}O + 6.4$. Superimposing the World Meteoric water Line (WML) on the same figure displays proximity with LML for the most enriched rainwater samples; however, the most depleted samples are located above the WML, and fit a rather Mediterranean trend. This LML is the same as that found in other areas of Morocco (Michelot 1991; Marah et al. 2007). It is generally attributed to coexistence, in this area, of rain from Atlantic and Mediterranean origins. Projecting in the same graph (Fig. 4) stable isotope ratios of springs emerging from the various Middle Atlas aquifers show that all of these springs are located around a trend line very close to that described above (LML), indicating that waters recharging these springs were infiltrated during a climatic period identical to that prevailing today, and that this infiltration is rapid, without stagnation favorable to evaporation. The one exception is spring Ain Regraga (sample No. 30) located below the WML, which seems to confirm the evaporation of its water prior to sampling. This assumption is verified by the location of this emergence within a swamp area; it is not a direct outlet of groundwater. Moreover, this figure shows also the heavy isotopes depleted water sampled versus altitude. Indeed, the folded Middle Atlas springs are the most impoverished, with values of δ^2H and δ^{18}O ranging between −46 and −60.50 ‰ versus SMOW, and −7.55 and −9.53 ‰ versus SMOW, respectively. The springs of the Middle Atlas limestone plateaus ('Causses') follow them in the graph, with values of δ^2H and δ^{18}O ranging between −37.51 and −51.64 ‰ versus SMOW, and between −6.16 and –8.04 ‰ versus SMOW, respectively.

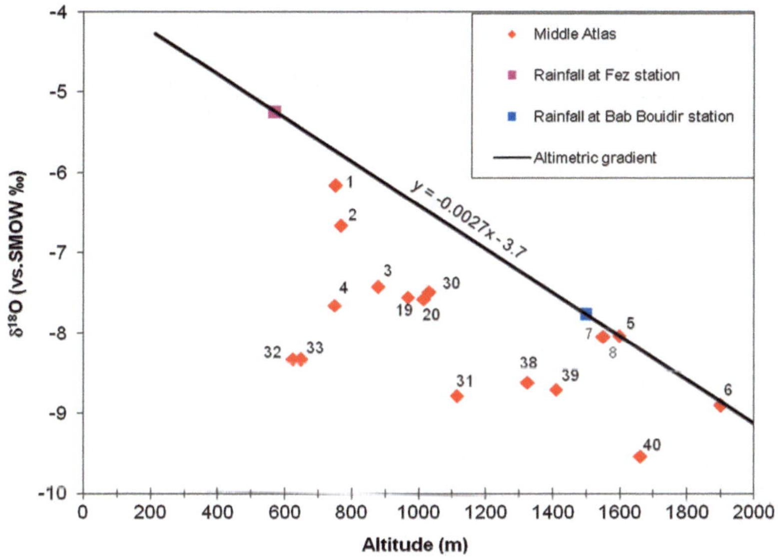

Fig. 4 Determining recharge altitude of the Middle Atlas groundwaters, using $\delta^{18}O$ altimetric gradient in rainwater

3.2 Stable Isotope Gradient Versus Elevation

Stable isotope gradient was determined by establishing a trend line that fits the springs with a well-known recharge area, with an average altitude close to that of the emergence zone. This trend reflects a gradient of 0.27 ‰ per 100 m gain in elevation (Sefrioui et al. 2010). Comparing this gradient with previous results obtained across Morocco can highlight similar results in the Rif mountains (ABHS 2002), in the Errachidia sedimentary Cretaceous basin (El Ouali 1999; El Ouali et al. 1999) and in the Tadla basin (Bouchaou et al. 1995). These results specify those obtained by Marcé (1975), who evaluated the gradient to -1 ‰ per 350 m, which corresponds to -0.286 ‰ per 100 m.

3.3 Average Elevation of Recharge Areas

Applying the trend equation established above to the Oxygen 18 content of each of the springs analyzed (Table 2), enabled determining the average altitude areas of the springs which are the purpose of the present study. Interpreting the results, in the light of local and regional hydrogeological data, can suggest the following

Table 2 Recharge altitudes of the main springs of the middle atlas, obtained from the gradient Oxygen-18 versus altitude

No.	Spring	Z (m)	$\delta\ ^{18}O$ (‰ vs. SMOW)	Z recharge (m)	Difference (m)
1	Ain Maarouf	752	−6.16	915	163
2	Ain Aghbal	769	−6.66	1,100	331
3	Ain Ribaa	880	−7.42	1,381	501
4	Ain Bittit	750	−7.65	1,467	717
5	Ain Zerrouka	1,597	−8.03	1,607	10
6	Ain Aberchane	1,900	−8.90	1,930	30
7	Ain Titzil	1,550	−8.04	1,611	61
8	Ain Sidi Rached	1,550	−8.04	1,611	61
19	El Anssar	968	−7.55	1,430	462
20	Ain Ras El Ma	1,015	−7.57	1,437	422
30	Ain Regrag	1,032	−7.48	1,404	372
31	Bouk	1,114	−8.78	1,885	771
32	Ain Sebou	625	−8.32	1,715	1,090
33	Timedrine	650	−8.33	1,718	1,068
38	Ain marmocha	1,324	−8.61	1,822	498
39	Skhounate S.Mdez	1,410	−8.70	1,855	445
40	Ain Tataw	1,660	−9.53	2,163	503

remarks: for springs of the Middle Atlas 'Causse' (limestone plateau), the differences in altitude between the recharge areas and the emergence zones vary from 10 m for the spring Zerrouka (No. 5), to 372 m for the spring Regraga (No. 30). This range appears to be related to the structure of the recharge areas drained by these springs. In fact, this difference is significant between the Zerrouka recharge area, which belongs to the more or less tabular 'Causse' of Ifrane (Bentayeb and Leclerc 1977), and the Ain Maarouf and to the Ain Regraga recharge areas, whose altitudes are increasingly higher. Although this difference in signature is the identical between the springs Ain Sidi Rached and Ain Titzil, respectively, it requires a different interpretation. Indeed, the first one is located on a (more or less) sloped and relatively small basin ('Causse' Ras El Ma), while the second is located in a basin largely tabular and flat, but of greater spatial extent ('Causse' Guigou).

These results corroborate those obtained by chemical analyzes, which emphasized a shift from the springs in which the altitudes are close to those of their recharge area (Zerrouka and Sidi Rached), to springs farther lower than their recharge area (Ain Regraga and AinTitzil).

For the folded Middle Atlas springs, a distinction can be highlighted between three clusters of springs. The first one includes spring No. 6, which recharge area's average altitude is close to its emergence altitude. It requires recharge from basaltic outcrops and/or from the Wadi Guigou's underflow. The second cluster consists of springs Nos. 20, 39, 19, 38, 40 and 3. These springs describe a stable isotope relationship, same slope as the regional gradient, with a greater depletion than rainwater, given their situation at higher elevations on the one hand, and subjected to a hydrogeological structure similar for the various compartments of the folded

Middle Atlas other. The third consists of the springs Nos. 32 and 33. These springs are those farthest from their recharge area, with differences in altitude of approximately 1,100 m, which justifies the large discharge of these springs.

4 Conclusion

Resolution of the issues raised by previous studies thus improves knowledge of the drained reservoirs, and helps in quantifying water resources for a more effective protection and a rational management of these resources.

Chemical characterization of these waters shows that their composition reflects that of the reservoir rocks they drain, on the one hand, and evolution they undergo during their flow through these reservoirs on the other hand.

The elevation gradient for Oxygen 18 has been calibrated at −0.27 ‰ per 100 m. The application of this gradient to stable isotope content of the sampled springs enables to infer average recharge altitudes for each of the springs. Comparison of these results to local and regional hydrogeological data shows a significant agreement for the springs whose watersheds were largely known, and provide a valuable guidance for the knowledge of springs for which the recharge area was poorly defined or totally unknown. These results contribute to rationally define the areas of the aquifer involved in the recharge of the springs, and helps in defining a strategy for the protection of this resource.

References

Agence du Bassin Hydraulique du Loukkous (ABHS) (2002) Etude isotopique de la dorsale calcaire du Rif—Internal report

Bentayeb A, Leclerc C (1977) Le Causse moyen-atlasique. In: 'Ressources en Eau du Maroc'—Tome III, Editions du Service Géologique du Maroc, pp 37–84

Bouchaou L, Michelot J-L, Chauve P, Mania J, Mudry J (1995) Apports des isotopes stables à l'étude des modalités d'alimentation des aquifères du Tadla (Maroc) sous climat semi-aride—C R Acad Sci Paris, série IIa 320:95–101

Chamayou J, Combe M, Genetier B, Leclerc C (1975) Le bassin de Meknès-Fès. In: 'Ressources en Eau du Maroc'—Tome II, Editions du Service Géologique du Maroc, pp 41–71

du Dresnay R, Studer G (1975) Carte géologique d'El Hajeb au 1/100.000°—Notes et mémoires No 160. Morocco Geol Surv

El Ouali A (1999) Modalités d'alimentation et échanges entre aquifères de piémont en conditions climatiques arides. Cas des systèmes aquifères du Haut Atlas/bassin crétacé d'Errachidia (Maroc)—'Etat'» Thesis, University Mohamed-V–School Mohammadia of Engineers, Rabat, 182 p

El Ouali A, Mudry J, Mania J, Chauve P, El Yamine N, Marzouk M (1999) Present recharge of an aquifer in a semi-arid region: an example from the Turonian limestones of the Errachidia bassin, Morocco Environ Geol 38(2):171–176

Essahlaoui A (2000) Contribution à la reconnaissance des formations aquifères dans le bassin de Meknès-Fès (Maroc). Prospection géoélectrique, étude hydrogéologique et inventaire des ressources en eau. Applied sciences thesis, School Mohammadia of Engineers, Rabat, Maroc, 258 p

Marah H, Zine N, Qurtobi M, Zerouali A (2007) Sens d'écoulement, vitesse et âge des eaux de l'aquifère turonien du bassin de Tadla (Maroc)—AJEAM-RAGEE 12:1–12
Marcé A (1975) Contribution des méthodes isotopiques à l'étude des modalités d'alimentation et de renouvellement des réserves souterraines du Maroc—report SGN. 447. LAB. Fr Geol Surv (BRGM), Orléans (France)
Michelot JL (1991) Hydrogéologie isotopique des systèmes aquifères de Fès-Meknès, Errachidia et Kheng-el-Hammam—Projet MOR/8/004, AIEA
Sefrioui S, Fassi Fihri O, El Ouali A, Marah H, Newman B, Essahlaoui A (2010) Utilisation des outils isotopiques pour la détermination de l'altitude de recharge des principales sources du bassin de Sebou—Maroc. J Geomaghreb 6

Detection of Underground Cavities by Combining Electrical Resistivity Imaging and Ground Penetrating Radar Surveys: A Case Study from Draa Douamis Area (North East of Algeria)

C. Fehdi, I. Nouioua, D. Belfar, L. Djabri and E. Salameh

Abstract A geophysical survey routine is proposed to detect underground cavities and dolines, based on the sequential application of ground penetrating radar (GPR) and electrical resistivity imaging or tomography (RESTOM). A case study near Cherea area (NE Algeria) demonstrates the applicability of these methods. Nine GPR profiles and two-dimensional RESTOM have been applied, with relative success, to locate paleo-collapses and cavities, and to detect and characterize karst at two sinkhole sites near Cheria city where limestone is covered by about 10 m of clayey soils. The survey results suggest that GPR and RESTOM are ideal geophysical tools to aid in the detection and monitoring of sinkholes and other subsurface cavities.

Keywords Resistivity tomography • RESTOM • GPR • Karst • Sinkhole • Cheria • Algeria

1 Introduction

Around Cheria there is a serious karstic hazard caused by Eocene Limestone dissolution; this results in subsidence and the collapse of Quaternary terraces and pediments producing dolines (Baali 2001). This active process is a major source

C. Fehdi (✉) · I. Nouioua · D. Belfar
Department of Geology, Cheikh El Arbi Tébessi University,
Route de Annaba N°71, 12000 Tébessa, Algeria
e-mail: fehdi@yahoo.fr

L. Djabri
Department of Geology, Annaba University, 23000 Annaba, Algeria

E. Salameh
Department of Geology, University of Jordan, Amman, Jordan

J. Mudry et al. (eds.), *H2Karst Research in Limestone Hydrogeology*,
Environmental Earth Sciences, DOI: 10.1007/978-3-319-06139-9_5,

of property damage and a potential cause of personal injuries. It is a growing problem because of rapid urban development and changes over recent decades in the manner in which town planning has controlled the area. While the population of Cheria has remained approximately constant, the city has multiplied its urban and industrial area by four times in four decades.

In many cases, karstic structures, such as subsidence features, voids and collapses, represent disruptions to the geometry of an originally near-horizontal layered system. Geophysical techniques can be used to identify the feature geometries by contrasts in the physical properties. These properties, such as density, magnetic susceptibility, electrical resistivity and conductivity, vary between the media involved, and materials, such as limestone, gypsum, siltstone, clay, sand, breccia, air and water, which all have different geophysical properties. Consequently, geophysical surveys are commonly applied for the detection of different types of subsoil anomalies. They allow large areas to be covered in short times and represent an efficient way to detect subsurface heterogeneities, including voids, subsidence areas or refilled cavities. Techniques used include seismic reflection and refraction (Cook 1965), gravimetry (Colley 1963; Butler 1984; Bishop et al. 1997; Rybakov et al. 2001), ground penetrating radar (Ballard 1983; Annan et al. 1991) and resistivity tomography (Zhou et al. 2002). More recently, surveying based on magnetic susceptibility differences has also proved useful for detecting and modeling cavities and soil heterogeneities on the outskirts of urban areas (Rybakov et al. 2005; Mathé et al. 2006).

Resistivity Tomography (RESTOM) is one of the most promising techniques for solving the sinkhole problem. RESTOM is a proven imaging technique where the theory and application are well documented in geophysical research literature (Griffiths and Barker 1993). Although RESTOM is a useful tool in mineral exploration (Sasaki and Matsuo 1993; Van Schoor and Duvenhage 2000), the technique is well suited to applications in the fields of hydrogeology, environmental science and engineering (Spies and Ellis 1995; Barker and Moore 1998). There have been many applications of electrical resistivity methods to detecting sinkholes and cavities, for example, see Burger (1992).

In this chapter, a case study is discussed to demonstrate the potential of RESTOM and GPR as routine geophysical tools for the monitoring of sinkholes and other subsurface cavities. The survey sites are located in M'chentel area near Cherea city (Fig. 1a, b and c).

2 Geological and Hydrogeological Setting

The Cheraa plain is located in north-east Algeria bounded by the Youkous to the north, the Thelidjen to the south and the El Ma Labiod basin to the east (Fig. 1a).

The Cheria Basin is a subsident basin, which was formed during the Miocene; it is a part of a narrow trough which forms a small portion of the great Mio-plio-quaternary tectonic depression of Cheria (Vila 1980). The basin is entirely filled by

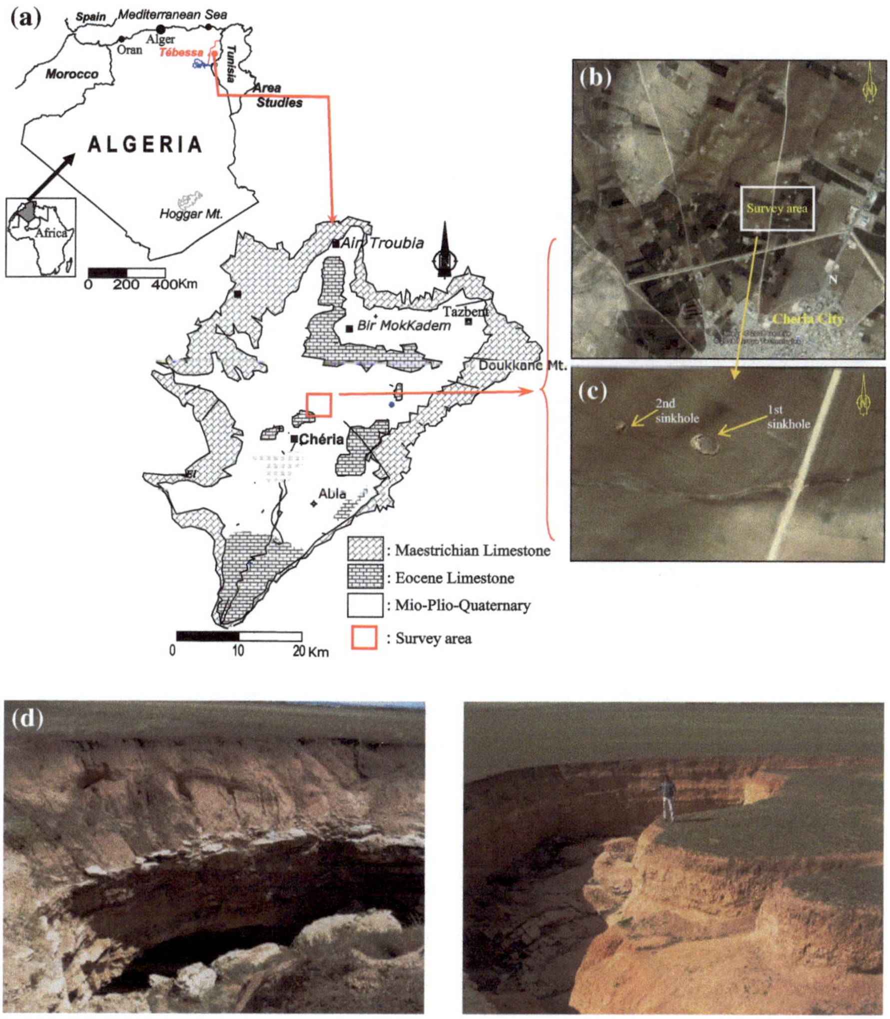

Fig. 1 Location of the study area. **a** Geological sketch of Cheria basin. **b**, **c** Satellite image of the study area (the *square* shows the studied area). **d** General view of the collapse. (Authors pictures. July, 2011)

marine sediments of Upper Cretaceous age. The bedrock is made up of marly rocks of the Danien Tertiary age and marly limestone rocks of Cretaceous age (Gaud 1977). The survey area investigated in the present study is from the Eocene formation, which is known locally as the Cheria Limestone formation. This formation is mainly made up of limestone which is overlain by Quaternary alluvial deposits, composed of gravel, sand, silt and clay.

From a hydrogeological point of view, Eocene limestone formation constitutes the most extensive aquifer in the Cheria Basin. The perennial water available in

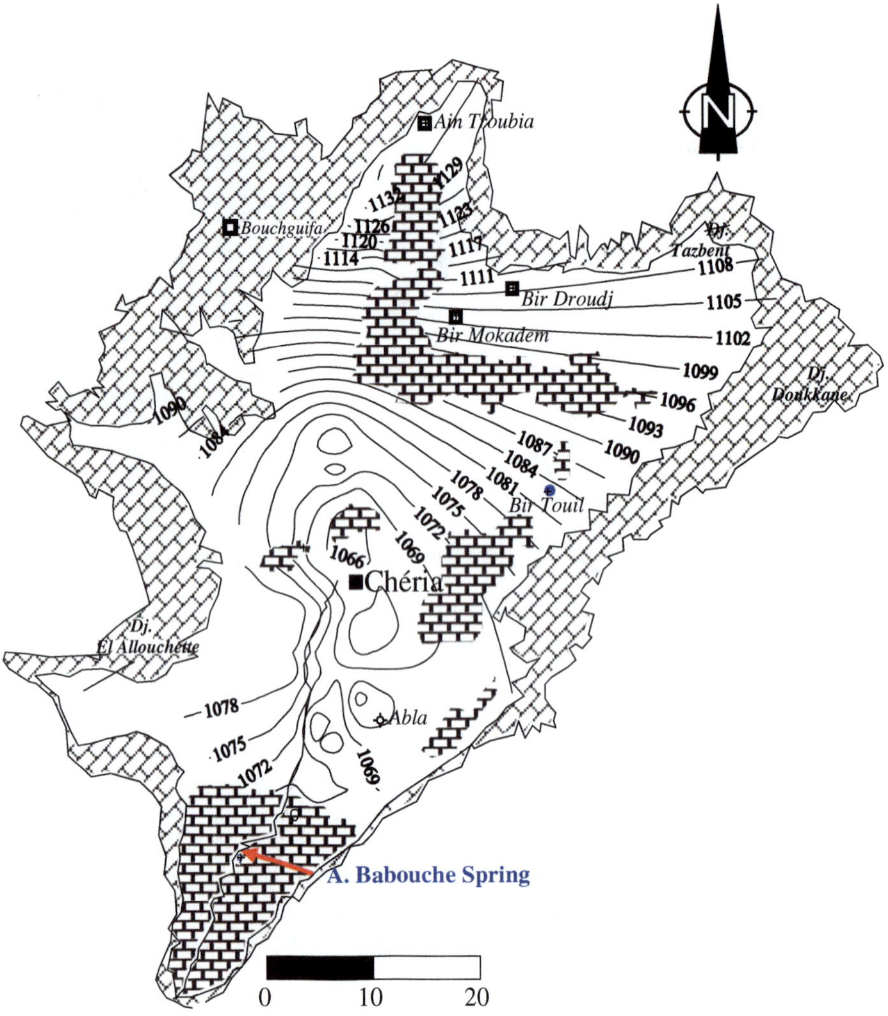

Fig. 2 Potentiometric map of Cherea plain (in Baali et al. 2007)

the study area is from the Eocene limestone aquifer. The Marly-limestone bedrock forms the boundaries of the groundwater reservoir (Figs. 2 and 3). The alluvial aquifer has not been appreciably utilized for water supply, because of the poor chemical quality of water, and its limited extent. The thickness of the Eocene limestone aquifer increases towards the central part of the basin (Chaffai et al. 2003).

Karst processes are one of the most important factors affecting the study area. The karst is generated by the dissolution of carbonates and, depending on its cohesive or non-cohesive behaviour, the subsequent collapse or dragging down of the overlying Quaternary cover. Seasonal oscillations in the water table are also an

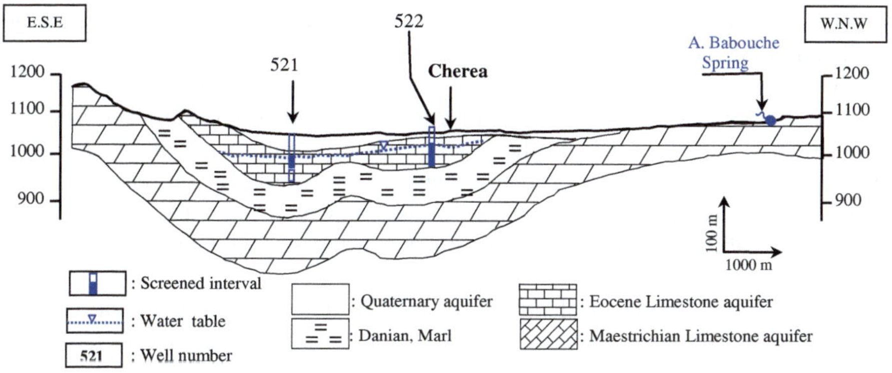

Fig. 3 Hydrogeological cross-section through Cherea plain (in Baali et al. 2007)

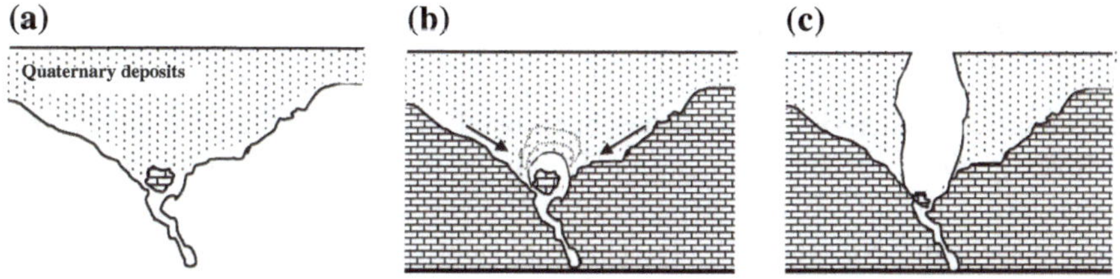

Fig. 4 Formation of a cover-collapse sinkhole. **a**, **b** and **c** represents different stage of sinkole formation

important factor in these processes. A result of this dissolution and collapse is the occurrence at outcrop of palaeocollapse features that vary in size and morphology (Baali et al. 2007). The recognized karst processes are both syn-sedimentary and post-sedimentary with the Quaternary deposits.

3 A Conceptual Model of Sinkhole Formation

A sinkhole results from either the transport of superficial material downwards along solutionally enlarged channels or collapse of the rock roof over a large bedrock cavity. Rock-collapse events are very rare; therefore, they constitute a minimal engineering hazard (Beck 1991). The more notable engineering hazard is created by the erosion processes in and beneath a soil cover. Of the sinkholes which occur in overburden, the cover-collapse sinkholes are the most unpredictable and are most likely to cause catastrophic damage to engineering works. The formation process of this type of sinkhole is shown in Fig. 4.

As shown in Fig. 4, the subcutaneous erosion starts with a buried solution sinkhole. The presence of a solution sinkhole is not a necessity; however, it certainly constitutes a potential area for a cover-collapse sinkhole because of its hydrologic properties (Williams 1985; Benson et al. 1998).

The erosion intensity near the base of the depression considerably exceeds that of the upper slopes as a result of concentrated runoff infiltration. The erosion process leads to a void in the overburden, especially when the materials of the overburden are cohesive. The void may grow upwards until it reaches a more resistant layer or the erosion process ceases, at which point the void attains a temporary stability. A sinkhole will form when the soil roof fails and collapses into the void.

The goal of a geophysical investigation is to delineate the potential collapse area when the formation process is still at stage B or C, where a void has developed but the soil has not yet collapsed.

4 Materials and Methods

The electrical resistivity method has been used in geotechnical and environmental investigations for about a century. It may be the most frequently used for site investigation in karst areas, especially when the overburden soil is clay-dominated (Cook and Nostrand 1954). The electrical conductivity of clayey soil and carbonate rock has an electrolytic origin, whereas most earth materials do not conduct electricity very well. According to Archie's law (LaMoreaux et al. 1984), electricity is conducted through interstitial water by ionic transport. Carbonate rock in general has a significantly higher resistivity than clayey soil because it has much smaller primary porosity and fewer interconnected pore spaces. Its typical resistivity value is more than 1,000 Ω m (Telford et al. 1990). Clayey materials tend to hold more moisture and have a higher concentration of ion to conduct electricity, therefore, have resistivity values <100 Ω m (Telford et al. 1990). The high contrast in resistivity values between carbonate rock and clayey soil favours the use of resistivity method to delineate the boundary between bedrock and overburden.

5 Field Investigation

5.1 Data Collection

Two-dimensional inversion techniques are becoming increasingly common and often satisfactory to assess the resolution and determine the limitations of the dataset, as shown by Dahlin (1996) and Dahlin and Loke (1998).

Electrical tomography profile was measured across the area, using SARIS Scintrex resistivity equipment by Wenner configuration with $a = 2$ m and 9 levels

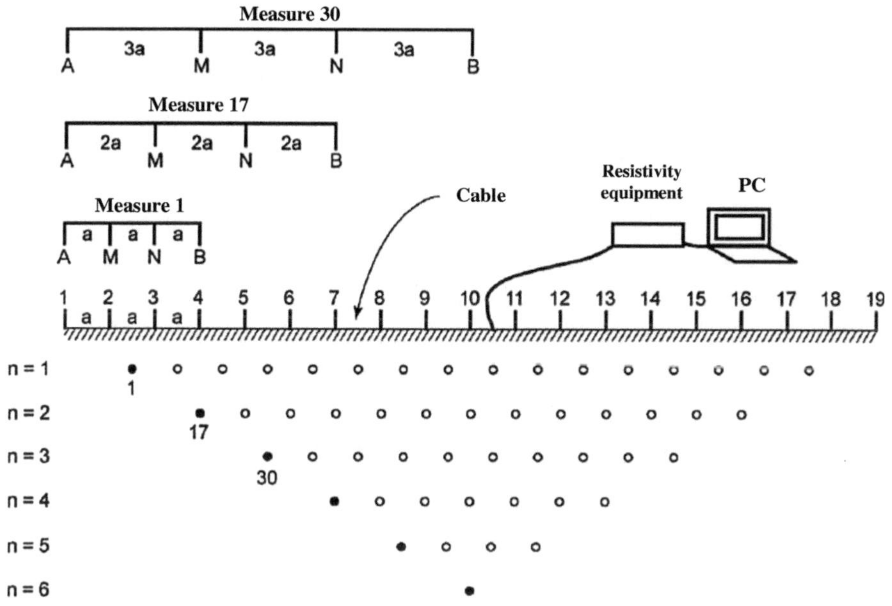

Fig. 5 Arrangement of electrodes for a 2D acquisition and measurement sequence for the Wenner configuration

of investigation, the total length of the profile through the collapsed site is 80 m. The survey traverses were oriented N-S. The method is based on measuring the electrical potentials between one electrode pair (M-N) while transmitting a direct current between another electrode pair (A-B; Fig. 5a). Figure 5b illustrates the pseudo-depth distribution of data points for the Wenner array configuration, used in this study.

GPR is a non-destructive geophysical method that produces vertical cross-sectional images of the shallow subsurface, the resulting image (radargram) being very similar in style to seismic reflection profiles. GPR acquisition is based upon the propagation, reflection and scattering of high frequency electromagnetic waves (generally ranging from 10 to 1,000 MHz) within the subsurface (Gutierrez et al. 2009).

Nine GPR profiles with a total length of 145 m, were conducted in the area enclosed by sinkholes 1 and 2. The GPR profiles were obtained using GSSI (U.S.A.) GPR with 200 and 400 MHz center frequency antennas.

Nine profiles were conducted; the first one was realized to determine behavior of signal and calibrate the appropriate depth (Fig. 6).

Four profiles have been realized for the first sinkhole; three of which are on the limestone roof and the fourth on the soil surface (Fig. 5). The rest of the GPR profiles were conducted in the areas west and north, near and around the second sinkhole (Fig. 7).

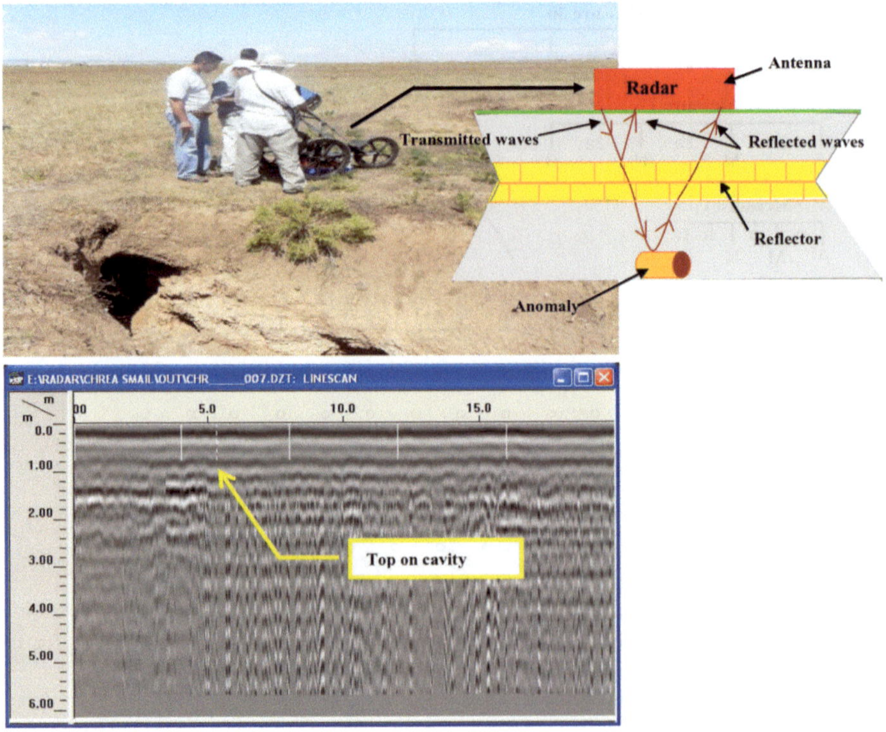

Fig. 6 Profile 1 with antenna 400 MHz: determination of behavior of signal and calibration investigation depth on the second sinkhole

Fig. 7 *Red arrow* profiles with antenna 400 MHz realized on the limestone roof of the first Sinkhole. *Green arrow* profile with antenna 400 Mhz realized on the soil surface

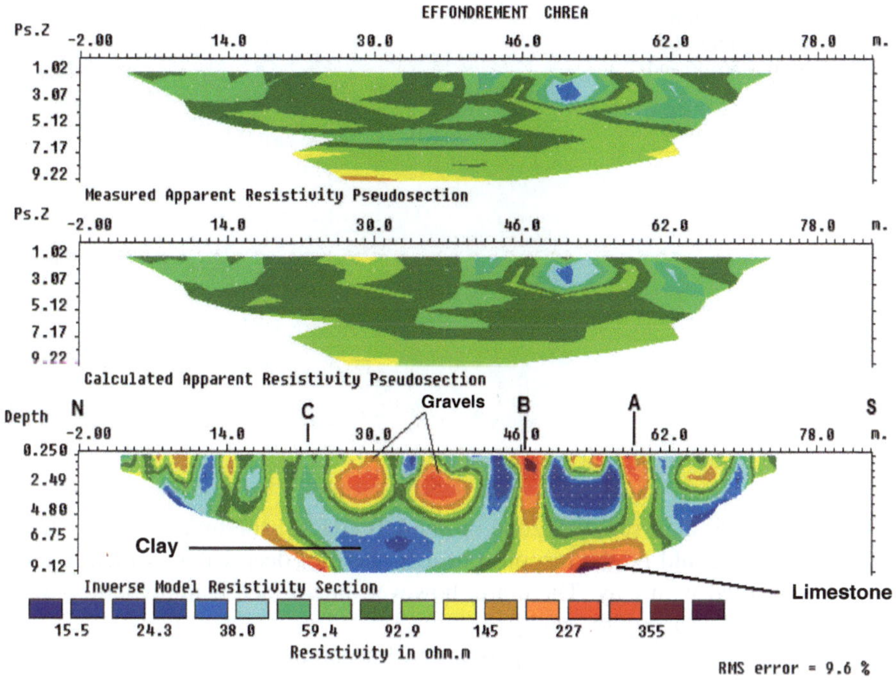

Fig. 8 The result of 2D inversion of Wenner array data from the study area (first Sinkhole)

6 Results and Discussion

6.1 Methods of Processing and Interpreting Data

6.1.1 Electrical Tomography

The resulting image of the Wenner array along the survey line, obtained over the study area (first sinkhole) (Fig. 8), is formed by three panels. The first panel, located on top, represents the pseudosection of the measured apparent resistivity. The middle pseudosection is the model calculated by the inversion program. The last panel is the inversion model section.

The image in Fig. 9 clearly maps the sinkhole structure that affected the study zone. The limits of three cavities can be seen as three smaller locally resistive zones A, B and C (at station positions 56, 46 and 22 m, respectively) within the basin-shaped zone.

These cavities are characterized by high resistivity values especially for the first and the second (A, B). A prominent, highly resistive feature (350–400 Ω m) is visible below station positions 56 and 46 m. This feature appears to have a depth extent of 3–6 m. Judging from the large resistivity contrast, this anomaly is the response of an air-filled cavity.

Fig. 9 Photograph showing the three cavities limits (at station positions 56, 46 and 22 m)

The relatively high resistive zone on the left edge of the image (below station position 22 m) could be the response of a further not deep cavity which is probably likely filled with clay. There are, however, no surface features indicative of the existence of a subsurface cavity.

The bedrock is resistant corresponding to Eocene limestone situated at a depth of 10 m. The resistivities of the limestone vary between 400 and 500 Ω m. The clays have a resistivity of about 10–40 Ω m. The two resistant peaks centered on the stations, respectively, 40 and 50 m of about 2.5 m thick represents a zone of highly resistant gravel.

6.1.2 Ground Penetrating Radar

The studied sinkhole area has been developed on a Plio-Quaternary deposit, locally <10 m thick, whose deposits have been affected by synsedimentary subsidence caused by the dissolution of the karstic bedrock (Fehdi 2011).

GPR surveys (200 and 400 MHz) were performed in the studied sector, a total number of five profiles around the second sinkhole. Unfortunately, on the set of these profiles data, cavities have not been detected. This non-detection is due to the heterogeneous nature of the soil (clay and wet soil) not suitable for this technique. GPR data for these profiles have shown in Fig. 10 an overall horizontal geometry.

Indeed, with the same equipment on the roof first collapsed limestone, which is more homogeneous and resistant soil, several anomalies have been detected that probably correspond to cavities (Figs. 10, 11 and 12).

The presence of a cavity is characterized by an amplification of the signal, and a reversal of the polarity when positioned on the cavity. Concerning fractures, the signal appears as a sharp and linear interface (Fig. 10).

The first profile PR/a shows a cavity located 10 m from the beginning of the profile and 0.5 m deep.

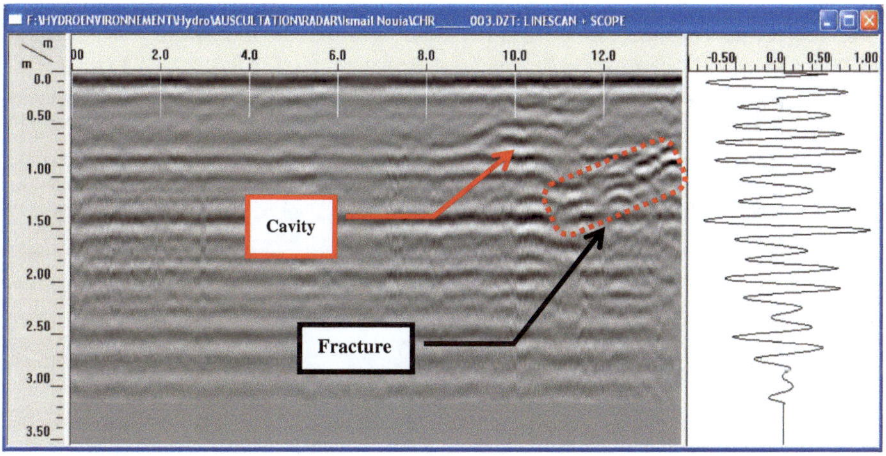

Fig. 10 Profile radargramm (PR/a) showing cavity and fracture

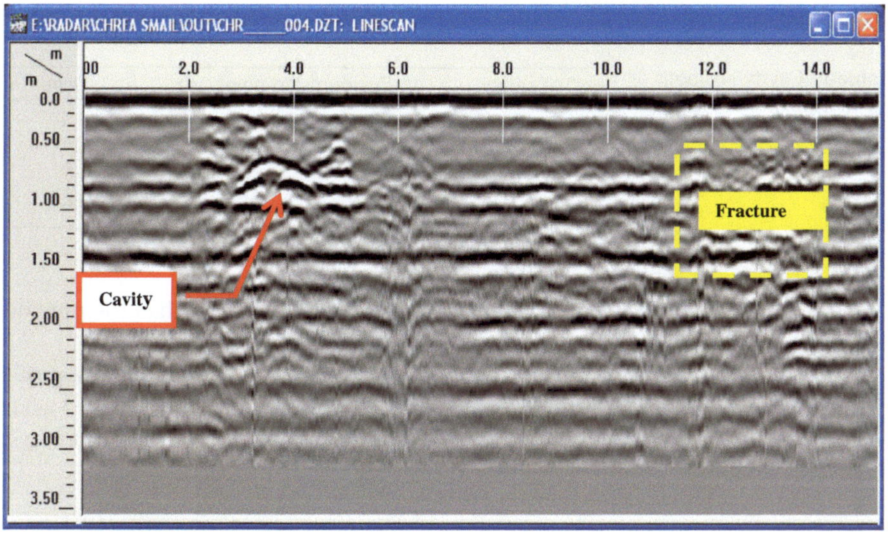

Fig. 11 Profile radargramm (PR/b)

The radargram (Fig. 9) also shows a signal decompression zone from 11 m to the end of the profile probably corresponding to a fracture with a depth ranging from 1.1 to 0.6 m.

The second profile PR/b (Fig. 10) shows a cavity at 2.2 m from the starting point and 50 cm deep.

At the end of the profile (between 11.7 and 14 m) a decompression zone probably corresponding to a small cavity or fracture to 0.7 m deep was found.

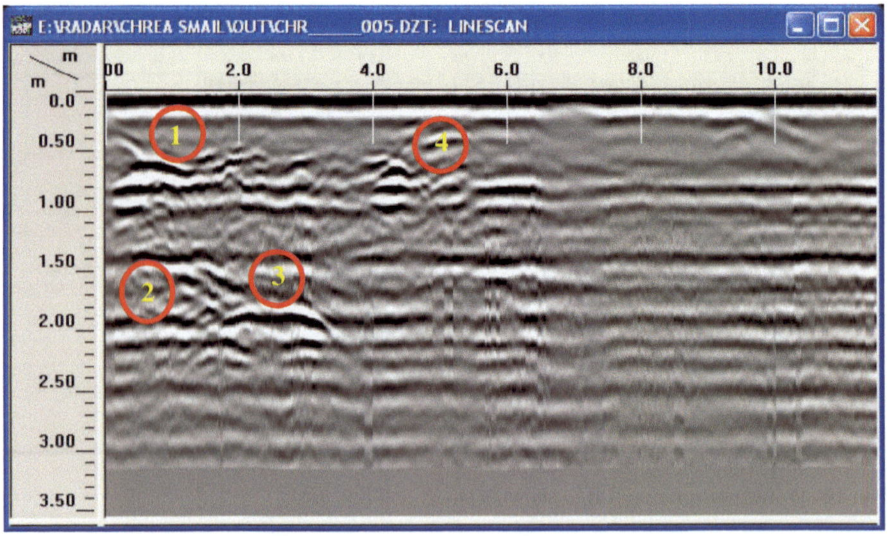

Fig. 12 Profile radargramm (PR/c)

Table 1 Cavity positions of the first sinkhole

N° of cavity	Position/start point (m)	Depth (m)
1	0.8	0.5
2	1.2	1.4
3	2.5	1.7
4	4.31	0.6

Moreover, the signal shows a field much more heterogeneous than the previous profile.

According to these results, a decision must be made to conduct another profile (PR/c) in cross-section with the previous profiles (PR/a and PR/b) (Table 1). The radargram (Fig. 12) shows a heterogeneous soil in the first 6 m with the presence of several cavities as resumed in Table 1.

7 Conclusions

The geophysical surveys have provided valuable information on the extent and geometry of the sinkhole structure. The high resolution of GPR prospecting allows the sub-soil structure to be characterized and the presence of hazards in deeper locations to be inferred. The authors show that although in many cases wetness and high clay contents of alluvial deposits are a handicap to the penetration depth of GPR-systems, its profiles provide an excellent image of the distribution and geometry of the deformation structures and sinkhole fill deposits located at shallow depth, especially in the homogeneous and resistant fields as limestone rock.

The electrical resistivity profiles, although with a smaller resolution, have provided valuable information on the subsidence structure and distribution of the stratigraphic units at greater depths.

GPR surveys (200 and 400 MHz) were performed in the studied sector, five profiles around the second sinkhole. Unfortunately, on the set of these profiles data, cavities have not been detected. This non-detection is due to the heterogeneous nature of the soil (clayey and wet soils) not suitable for this technique, whereas with the same equipment on the roof of the first collapsed limestone, which is more homogeneous and resistant soil, several anomalies have been detected that probably correspond to cavities and fractures which are located at different depths.

The 2D cross-borehole resistivity tomographies conducted in this study allowed mapping the complex geometry of the karstic subsurface conditions. The tomographic images computed within Res2DInv reveal the presence of three cavities' limits, which can be seen as three smaller locally resistive zones A, B, and C (at station positions 56, 46, and 22 m, respectively) within the basin-shaped zone.

These cavities are characterized by high resistivity values, especially for the first and the second (A, B). A prominent, highly resistive feature (350–400 Ω m) is visible below station positions 56 and 46 m. This feature appears to have a depth extent of 3–6 m, judging from the large resistivity contrast; this anomaly is the response of an air-filled cavity.

The relatively high resistive zone on the left edge of the image (below station position 22 m) could be the response of a further, not deep, cavity, which is probably likely filled with clay.

References

Annan AP, Cosway SW, Redman JD (1991) Water table detection with ground-penetrating radar. In: Society of exploration geophysicists (annual international meeting program with abstracts), pp 494–497

Baali F (2001) Eude hydrogéologique hydrochimique de la région karstique de Chéria N-E Algérien. Magister Univ Annaba Algérie, p 100

Baali F, Rouabhia A, Kherici N, Djabri L, Bouchaou L, Hani A (2007) Qualité des eaux souterraines et risque de pollution en milieu semi aride. Cas de la cuvette de Chéria. NE algérien. J Estudios Geologicos 63: CSIC, Spain

Ballard RF (1983) Cavity detection and delineation research. Report 5, Electromagnetic (radar) techniques applied to cavity detection. Technical Report GL, 83–1, p 90

Barker R, Moore J (1998) The application of time-lapse electrical tomography in groundwater studies. Lead Edge 17(10):1454–1458

Beck BF (1991) On calculating the risk of sinkhole collapse. In: Kastning EH, Kastning KM (eds) Proceedings of Appalachian Karst Symp, Radford, Virginia, 23–26 March 1991. National Speleological Society, Huntsville, Alabama, pp 231–236

Benson RC, Kaufmann RD, Yuhr LB, Martin D (1998) Assessment, prediction and remediation of karst conditions on I-70, Frederick, Maryland, 49th Highway Geology Symposium, Precott, Arizona, 10–14 September 1998. Arizona Department of Transportation, Material Group, Arizona Geological Survey, pp 313–325

Bishop I, Styles P, Emsley SJ, Ferguson NS (1997) The detection of cavities using the microgravity technique: case histories from mining and karstic environments. Geol Soc Eng Geol Spec Publ 12:153–166

Butler DK (1984) Microgravimetric and gravity gradient techniques for detection of subsurface cavities. Geophysics 49(7):1084–1096

Burger HR (1992) Exploration geophysics of the shallow subsurface. Prentice-Hall, Englewood Cliffs, p 489

Chaffai H, Baali F, Djabri L, Rouabhia Aek (2003) Facteurs influençant le chimisme des eaux dans une zone semi-aride: Cas des aquifères d'El Ma El Abiod, Tébessa, Hammamet et Chéria. ICOWAP-Sep 2003. Colloque Oasis, Eau et population, Biskra Algérie, pp 339–344

Colley GC (1963) The detection of caves by gravity measurements. Geophys Prospect XI:1–9

Cook JC (1965) Seismic mapping of underground cavities using reflection amplitudes. Geophysics 30(4):527–538

Cook KL, Van Nostrand RG (1954) Interpretation of resistivity data over filled sinks. Geophys Prospect 21:716–723

Dahlin T (1996) 2D resistivity surveying for environmental and engineering applications. First Break 14:275–284

Dahlin T, Loke M (1998) Resolution of 2-D Wenner resistivity imaging as assessed by numerical modelling. J Appl Geophys 38:237–249

Fehdi Ch, Baali F, Boubaya D, Rouabhia A (2011) Detection of sinkholes using 2D electrical resistivity imaging in the Cheria Basin (North-East of Algeria). Arab J Geosci 4:181–187

Gaud J (1977) Etude géologique et hydrogéologique du plateau de Chéria Wilaya de Tébessa. Rapport interne No. 2. A.N.R.H de Tébessa (Agence Nationale des ressources hydriques), p 96

Griffiths DH, Barker RD (1993) Two-dimensional resistivity imaging and modeling in areas of complex geology. J Appl Geophys 29:211–226

Gutierrez A, Galve JP, Lucha P, Bonachea J, Jorda L, Jorda R (2009) Investigation of a large collapse sinkhole affecting a multi-storey building by means of geophysics and the trenching technique (Zaragoza city, NE Spain). Environ Geol 58:1107–1122

LaMoreaux PE, Wilson BM, Memon BA (1984) Guide to the hydrology of carbonate rocks. UNESCO, France

Mathé V, Léveque F, Mathé PE, Chevallier C, Pons Y (2006) Soil anomaly mapping using a caesium magnetometer: limits in the low magnetic amplitude case. J Appl Geophys 58:202–217

Rybakov M, Goldshmidt V, Fleischer L, Rotstein Y (2001) Cave detection and 4-D monitoring: a microgravity case history near the Dead Sea. Lead Edge (Soc Explor Geophys) 20(8):896–900

Rybakov M, Rotstein Y, Shirman B, Al-Zoubi A (2005) Cave detection near the Dead Sea-a micromagnetic feasibility study. The Leading Edge (Society of Exploration Geophysicists) 24(6):585–590

Sasaki Y, Matsuo K (1993) Surface-to-tunnel resistivity tomography at the Kamaishi Mine. Butsuri-Tansa 46:128–133

Spies B, Ellis R (1995) Cross-borehole resistivity tomography of a pilot sale, in situ vitrification test. Geophysics 60:886–898

Telford WM, Geldart LP, Sheriff RE (1990) Applied geophysics, 2nd edn. Cambridge University Press, New York

Van Schoor M, Duvenhage D (2000) Comparison of crosshole radio imaging and electrical resistivity tomography for mapping out disseminated sulphide mineralisation at a surface test site in Mpumalanga, South Africa. Explor Geophys 30:3–4

Vila JM (1980) La chaîne alpine de l'Algérie orientale et des confins Algéro-Tunisiens. Thèse de Doctorat- es-sciences, Université Pierre et Marie curie, Paris VI

Williams PW (1985) Subcutaneous hydrology and the development of doline and cockpit karst. Z Geomorphol NF 29(4):463–482

Zhou W, Beck BF, Adams AL (2002) Effective electrode array in mapping karst hazards in electrical resistivity tomography. Environ Geol 42:922–928

Using Hybrid Genetic Algorithms in Assembling Master Recession Curves of Karst Springs

Miloš Gregor and Peter Malík

Abstract Computation methods, using the evolution algorithm concept—mainly group of optimization methods known as genetic algorithms—provide an interesting methodical approach, applicable for a construction (not interpretation) of assembled recession curves. Assemblage of the master recession curve from various hydrograph segments by the help of the hybridized genetic algorithm approach can avoid obstacles such as limited time-series datasets, incomplete recessions or too many segments in many recession curves, complicated hydrograph shape in the case of karstic springs (caused by combined laminar and turbulent discharge sub-regimes due to karst network settings), different time intervals of observations (daily or weekly frequencies), short time-series intervals, imprecise measurements, different types of datasets (averaged or directly measured data) or even rough (inaccurate) measurements of discharges. The only type of discharge curve without practical output solution of the method presented is the constant outflow. In practice, assembling of recession curves is necessary for hydrogeological phenomena, especially in moderate climate regions, where recessional periods are frequently interrupted by partial recharge events. The presented algorithm was already implemented to a programme solution, so that the applicability of the hybrid genetic algorithms method for master recession curve creation is at hand without requirements on programming skills of hydrologists involved, which can be considered as an immediate advantage of the method described.

Keywords Hydrogeological structure • Hydrological drought • Groundwater vulnerability • Master recession curve assembling

M. Gregor · P. Malík (✉)
Department of Hydrogeology and Geothermal Energy, SGUDS—Geological Survey of Slovak Republic, Mlynská dolina 1, 817 04 Bratislava 11, Slovakia
e-mail: peter.malik@geology.sk

M. Gregor
e-mail: milos.gregor@geology.sk

J. Mudry et al. (eds.), *H2Karst Research in Limestone Hydrogeology*, Environmental Earth Sciences, DOI: 10.1007/978-3-319-06139-9_6,

1 Introduction

Runoff from a catchment (or hydrogeological structure) is physically determined by two types of impacts and parameters. In the first influencing category, there are static properties of the environment (as e.g. geometry, hydraulic properties, catchment slopes), which remain constant over time. In the second influencing category, dynamic properties are counted, such as degree of structure (rock and soil mass) repletion by water, rate of the water infiltration, rainfall and its variation in time and space, evapotranspiration, possible anthropogenic influence and other time-dependent factors.

Analysis of recession curves reflecting runoff from a catchment or discharge of spring is often the method used in hydrological practice, which provides a wide range of result interpretations. Although the principles of hydrograph analyses have been developed for more than one and a half of a century (Boussinesq 1877, 1904; Maillet 1905; Horton 1933; Barnes 1939; Cooper and Rorabaugh 1963; Schoeller 1965; Drogue 1967; Kullman 1980, 1990; Padilla et al. 1994; Griffiths and Clausen 1997; Kovács 2003; Gregor 2008), the introduction of computer techniques mainly enhanced their applications in the broader field of hydrology and hydrogeology. Such alternative applications can be found, for example, in the analysis of hydrological drought (Tallaksen and van Lanen 2004), groundwater vulnerability (Kullman 2000; Malík 2007) and in the separation of runoff components, e.g. baseflow separation (Malík 2010). Several optimization techniques, tending to solve many of the shortcomings and gaps of the existing methods, were already introduced to hydrogeological and hydrological practice (Lamb and Beven 1997; Rutledge 1998; Posavec et al. 2006).

In rough approach, it may look like the recession curves have generally a similar course; however, the combination of the aforementioned parameters causes each recession process to be unique and differs in absolute values of discharge sequences. The influence of the dynamic parameters described is projected on the complexity of recession curves described within even a simple watershed and represents the main problem to be solved in the process of master recession curve assembling.

The second problem is found in a considerable variability in runoff, which means that it is usually impossible to mathematically describe a recession curve only by the use of one simple regression equation, and it is necessary to use a set of multiple staked regression equations.

The third problem is that in many cases it is impossible to describe the runoff from different catchments and hydrogeological structures only by the use of one type of equation. Especially in the cases with flux type change (e.g. from turbulent to laminar; Kullman 1983, 1990), it is necessary to find several matching equation types.

The last task is to overcome the problem with the accuracy of discharge measurements and the frequency of measurements. Often, especially at low discharges, the visible change between two readings in time is too small to be recorded. On the other hand, if the typical flood wave duration in comparison with the frequency of readings is too short, the probability of maximal discharge recording is limited. This may represent a serious deficiency, especially in the case of groundwater monitoring, where many objects (springs or wells) are monitored only on weekly bases, but a typical flood-wave peak after a storm event is shorter. In many cases, especially when discussing specific operational monitoring objects, only short discharge time-series (e.g. not longer for more than two years of observations) are available. It is very probable that within such a limited time of observations the extreme values are omitted from the available set of results.

Over time, several methods have been developed for this purpose. The most simple and the first in the process of development is a manual creation of a master recession curve. Using this method, all recession curves are shown on a single chart. In the process of assemblage, each recession curve is moved on the horizontal timeline until the individual curves will form one, more or less compact, master recession curve. The main deficiency of this method is the time-consuming processing and also strong dependency of results on the authors' influence. Usually, different authors obtain different recession results, depending on their previous experience.

2 Materials and Methods

Genetic algorithms are, together with neural networks, expert systems and methods of chaos theory, considered to be a part of an artificial intelligence group of methods. They are based on the principles of Darwin's evolutionary theory and can be described as stochastic search methods, based on the mechanism of natural selection and on the principles of genetics. In the field of water science, these were used in parallel evolutionary calibration of hydrological models (Sharma et al. 2006); determination of rock hydraulic parameters from the pumping tests (Samuel and Jha 2003); optimizing of water resources abstraction (Chiu et al. 2007) or contamination transport modelling (Bayer and Finkel 2004).

The main principle applied within genetic algorithm procedure is the creation of initial population of random solutions of the defined problem at the beginning. Each solution in the population is presented as data structure (e.g. data array), which allows us to effectively save the information about individual solution. Individual solutions are tested and evaluated towards achieving a defined goal. Subsequently, from the existing population of solutions, a new population is created, while better solutions from the previous population are equipped with higher probability of transition into a new generation in comparison to solutions with lower valuation. In the creation process of a new generation of solutions, two individual solutions are randomly selected; however, solution with better rating has a higher probability to be selected. For this purpose, a "roulette mechanism" is frequently used, where solutions with a higher rating obtain a bigger slice from the roundel. After defined crossing is performed, a newly created solution takes the properties from both of the two solutions from the previous generation. This

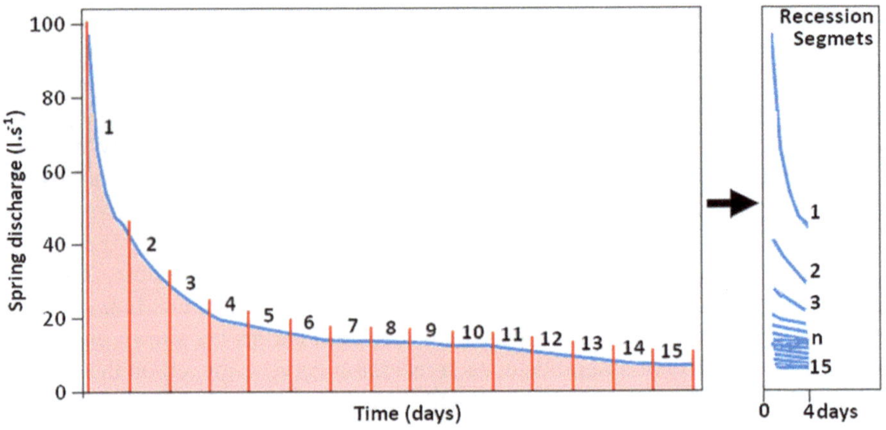

Fig. 1 Division of selected depleting sections of the hydrograph into N-days segments

new individual solution presents a random combination of the parental generation properties. During the crossing process, a random mutation which randomly changes a part of the final solution also occurs. This mutation process, however, occurs only with very low probability.

Gradually, with help of evolution cycles repetition, a solution that is very close to the ideal one is evolved. The described evolutionary process can be stopped either by a pre-described number of evolutionary cycles in the set, or when a sufficiently accurate solution is reached, or even if the population losses its ability to converge to the ideal solution. Prior to the creation of a master recession curve, a detailed study of the hydrogram should be performed. After individual segments of recession had been chosen, each of the selected intervals of recession is divided to a corresponding number of individual N-day segments (4-days long segments were found to be most useful). By segmentation, the method can overcome the "shading influence" of the dynamic factors (e.g. recharge and runoff variability). These segments are later used in the assemblage process to create the master recession curves (MRC). Ideally, resulting MRC is assembled by the simple moving of the defined short segments on the timeline until they will together create one single line. In practice, this process is not so simple and for the automatic shifting of separate segments on the timeline, a described genetic algorithm is used.

3 Results

In the process of MRC assemblage from measured data, the structure of each separate solution in the population of solutions had to be defined, as shown on Fig. 1. After splitting of selected discharge recession intervals, several hundred segments were usually obtained (Fig. 1). Each individual segment was the member of the population of solutions, and in the population of solutions was defined by two-dimensional array

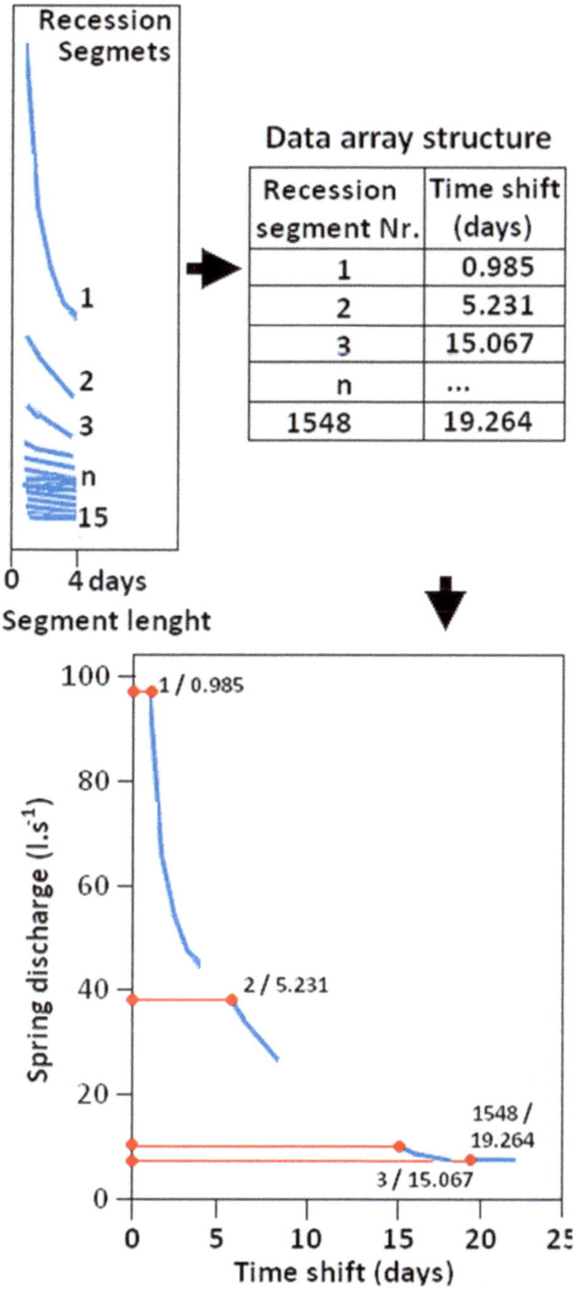

Fig. 2 Definition of data structure for each individual solution in the population of solutions

(Table; Fig. 2). Depending on the number of pre-defined solutions in the population, a corresponding number of data structures were created. In these, individual solutions were defined. The first population was then created pseudo-randomly.

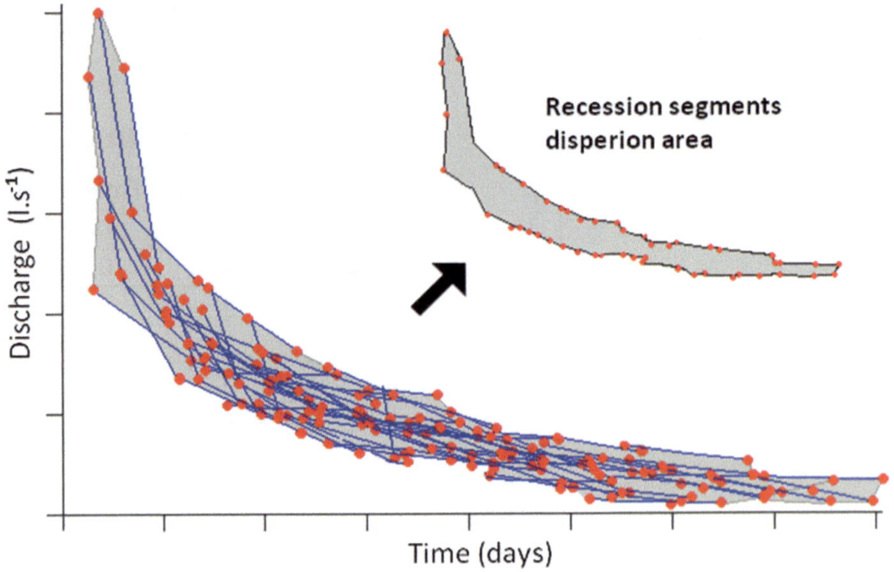

Fig. 3 Definition of segments' dispersion area

The maximal time shift tmax was pre-defined before the run of the algorithm—and the movement of segments in the evolution process automatically did not overstep this value. A random shift between the initial time t_0 and maximal time t_{max} was then generated for each segment. Initial time t_0 here represents the peak (maximum) discharge in the time–series dataset.

In the process of random shift generating for the initial population, the range of possible random time shift values increases with decreasing values of discharges in segments. After the initial—first generation had been created, individual solutions were assessed from the "best-fit" point of view. As an assessment tool for the accuracy of solutions, the dispersion area of the segments (Fig. 3) was calculated for each solution.

Following the assessment of solutions in one generation, solutions for the next generation were then prepared. To do this, two solutions are randomly selected from the current generation by the use of "roulette selection mechanism", where better solutions are equipped with higher probability to be used as a source for creating the solutions in the next generation. After the two solutions from the current generation of solutions have been selected, only one solution is created for the next generation by the crossing process from the two ones. In this case, a method of "crossing by averaging" (Fig. 4) was applied. Here, a new individual solution copies the data structure, while the time shift for each segment is calculated as an average of the previous two solutions. In addition to the crossing process, also the mutation processes (Fig. 5) are incorporated in the evolution.

Usually the mutation occurs with very low probability (e.g. 1 mutation per 1,000 crossings); however, in this case, this approach was modified—hybridized,

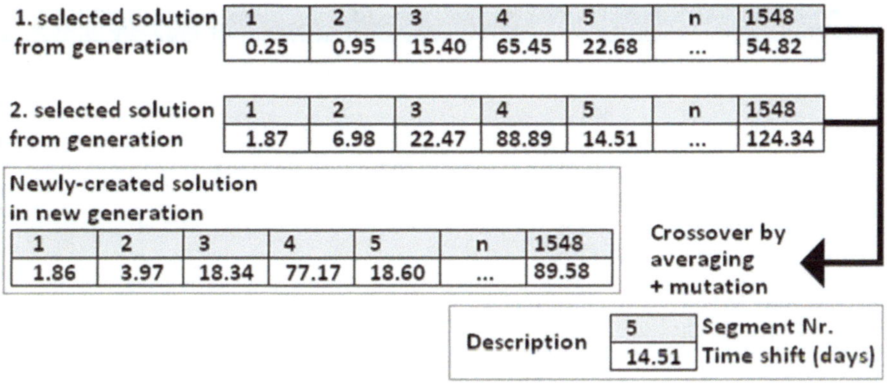

Fig. 4 Description of "crossing by averaging" method, used for creating of the new solution

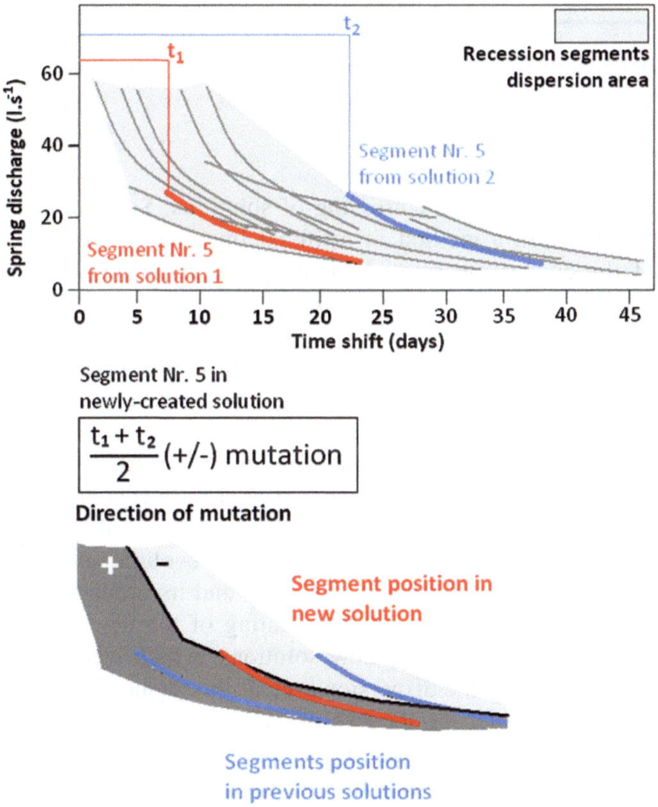

Fig. 5 Description of mutation process

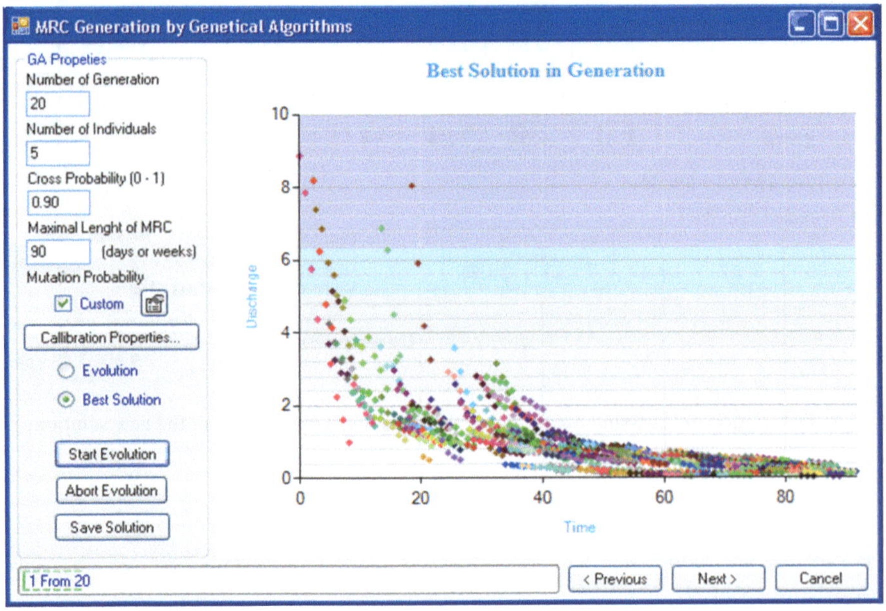

Fig. 6 Monitoring possibilities of evolution process by displaying of best solution in actual generation

as the mutation occurs at every crossing of solutions. Such an approach is frequently applied within a method of artificial immune system. This allows the reduction of the conversion rate of calculation and the population during the evolution remains relatively variable. Mutation is applied after the averaging of two solutions. Mutation value is just a random value generated from the defined interval $<t_0, t_{def}>$. If this range of random values were chosen to be too low, also the variability of the populations would be relatively low. Conversely, if the range was chosen to be too wide, after several cycles of evolution the development of the ultimate solution would stop, as the mutation would not permit the reduction of the value of the dispersion area segments. Therefore, the predefined range of values, from which random mutation values can be created, gradually decreases with the development of the evolutional cycles. It is also possible to set the maximal range of mutations and its gradual reduction within the run of algorithm application on the creating of the new solution. The procedure gradually created new and new solutions, which were continuously reassessed by calculation of the dispersion area of the segments.

Monitoring possibilities of the evolution process are either by displaying of the best solution in actual generation (Fig. 6) or by displaying of dispersion area change in the average solution and the best solution in successive generations.

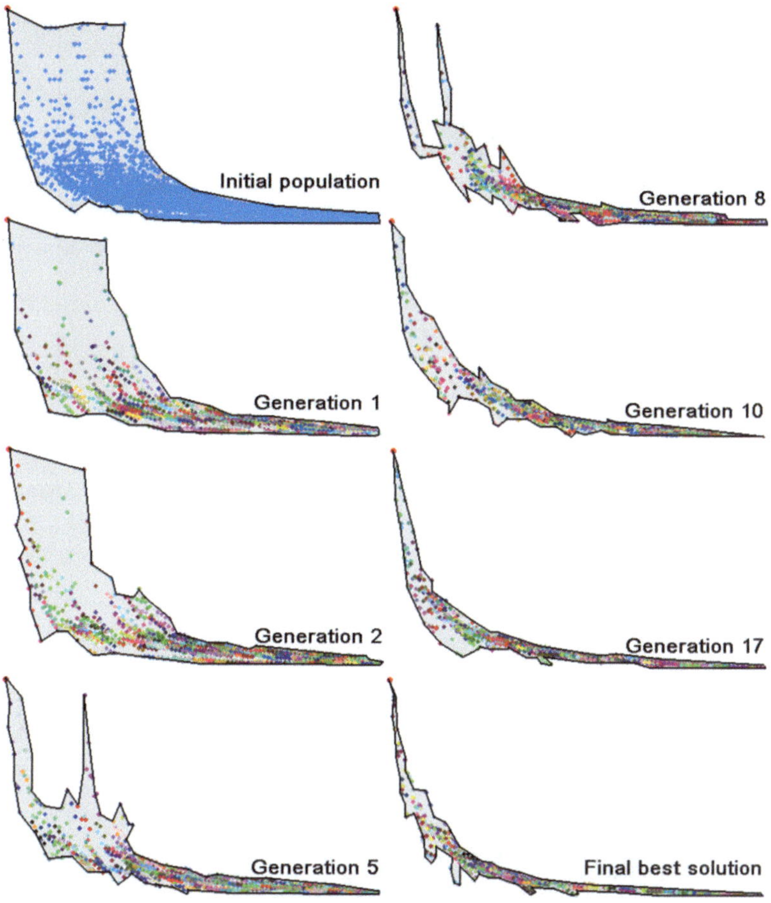

Fig. 7 Example of gradual evolutionary assembling of master recession curve (MRC) by the use of strongly hybridized genetic algorithm

The evolutionary cycle was repeated until the number of its runs reaches a predefined number of cycles. Application of this procedure gradually leads to creation of the solution corresponding to the resulting MRC from the initial pseudo-random generation (Fig. 7).

4 Discussions

After its implementation into the program, the described hybridized algorithm was tested and used at many very variable conditions. The result of MRC evolution from the recession curves selected from spring yield time series of a karstic spring Dovalka in Muráň (Central Slovakia; 48.7403°N; 20.0450°E) is shown on Fig. 8.

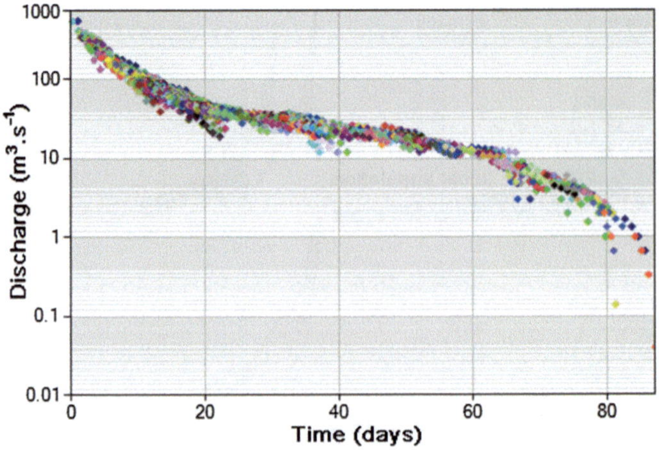

Fig. 8 Result of master recession curve assembling for Dovalka karst spring in Muráň (Central Slovakia)

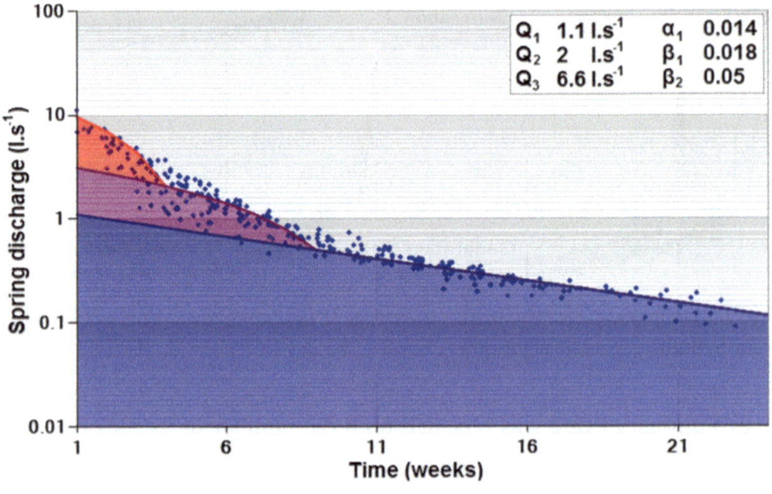

Fig. 9 Result of master recession curve assembling for Pod Mravinami spring in Súľov (NW Slovakia)—logarithmical plot

Three runoff sub-regimes of turbulent flow type are visible on the course of recession. The second example (Figs. 9 and 10) shows the result of a MRC assemblage from the spring discharge measurements of Pod Mravinami spring in Súľov (NW Slovakia; 49.151°N; 18.589°E). In the resulting MRC, two runoff sub-regimes with laminar flow type, plus additional turbulent sub-regime was identified. These are plotted on a logarithmical plot on Fig. 9, while Fig. 10 represents a normal plot view on a typical MRC of Pod Mravinami spring.

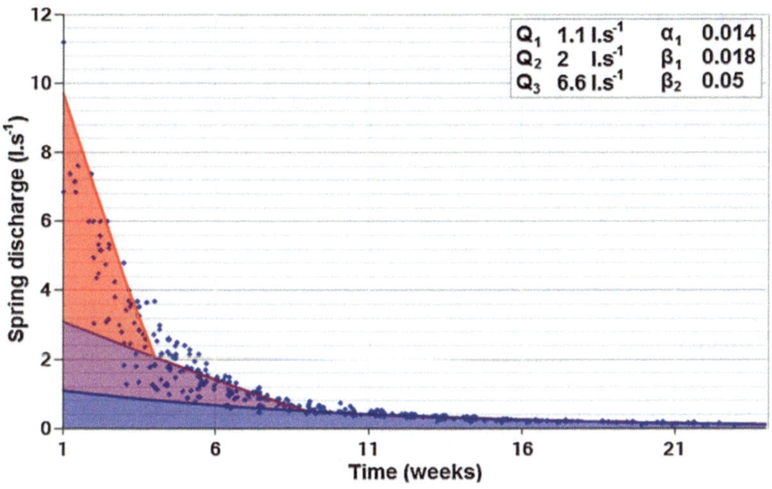

Fig. 10 Result of master recession curve assembling for Pod Mravinami spring in Súľov (NW Slovakia)—normal plot

5 Conclusions

The use of strongly hybridized genetic algorithm allows the creation of master recession curves at very variable conditions. This can avoid obstacles such as limited time-series datasets, incomplete recessions or too many segments in many recession curves. In the case of karstic springs, usage of strongly hybridized genetic algorithm helps to overcome problems with complicated hydrograph shape, caused by combined laminar and turbulent discharge sub-regimes due to karst network settings. It is also helpful if different time intervals of observations (daily or weekly frequencies), short time–series intervals, imprecise measurements, different types of datasets (averaged or directly measured data) or even rough (inaccurate) measurements of discharges are found in the time series analyzed. The only type of discharge curve without practical output solution of the method presented is the constant discharge value in time, which under real hydrological circumstances, however, in natural conditions never occurs over a longer period of time. The described algorithm was implemented into a programme solution, and it is possible to try the evolutional assembling of the master recession curve within the RC 4.0 module in the freely accessible HydroOffice software (http://www.hydrooffice.org), developed by one of the authors.

When master recession curve is completed—manually or with the help of the hybridized genetic algorithm—it is possible to apply any type of mathematical recessional models (linear, exponential, hyperbolical, power model or their combination) for further interpretation. Description of recession functions, as well as for deriving information on the physical functioning of the hydrogeological systems is a subject of additional studies, additional to the assemblage of MRC by the use of hybrid genetic algorithms.

Acknowledgments The authors thank the Agency of the Ministry of Education, Science, Research and Sport of the Slovak Republic for the Structural Funds of EU (ASFEU), providing assistance for the implementation process of EU Structural Funds, additional tools and sources for both field studies in the area and discharge time series analyses were accessible within the framework of the project Research on the climate change impact on the available groundwater amounts in Slovak Republic and the creation of expert GIS (ITMS Code No. 26220220002, Operational Programme Research and Development), which is hereby deeply acknowledged.

References

Barnes BS (1939) The structure of discharge-recession curves. Trans Am Geophys Union 20(4):721–725. doi:10.1029/2003WR002675

Bayer P, Finkel M (2004) Evolutionary algorithms for the optimization of advective control of contaminated aquifer zones. Water Resour Res 40(6):19 p. doi:10.1029/2003WR002675

Boussinesq MJ (1877) Essai sur la theories des eax courantes. Memoires presentes par divers savants a l'Academie des Sciences de l'Institut National de France, Tome XXIII (1)

Boussinesq MJ (1904) Recherches théoriques sur l'écoulement des nappes d'eau infiltrées dans le sol et sur le debit des sources. Journal de mathématiques 5e(10)

Chiu YCh, Chang LCh, Chang FJ (2007) Using a hybrid genetic algorithm—simulated annealing algorithm for fuzzy programming of reservoir operation. Hydrol Process 21(23):3162–3172

Cooper HH Jr, Rorabaugh MI (1963) Groundwater movements and bank storage due to flood stages in surface streams. USGS water supply paper (1536-J): 343–366

Drogue C (1967) Essai de détermination des composantes de l'écoulemet des sources karstiques. Evaluation de la capacité de rétention par chenaux et fissures. Bureau de Recherches Geologiques et Miniéres, Paris. Chronique d'Hydrogéologie (10)

Gregor M (2008) Vývoj programov na analýzu časových radov výdatností prameňov a prietokov vodných tokov [Software development for time-series analysis of springs yields and river discharges; in Slovak]. Podzemná voda 14(2):191–200

Griffiths GA, Clausen B (1997) Streamflow recession in basins with multiple water storages. J Hydrol 190:60–74

Horton RE (1933) The role of infiltration in the hydrological cycle. Trans Am Geophys Union 14:446–460

Kovács A (2003) Geometry and hydraulic parameters of karst aquifers—a hydrodynamic modelling approach. PhD. thesis, La Faculté des sciences de l'Université de Neuchâtel, Suisse, pp 131

Kullman E (1980) L'evaluation du regime des eaux souterraines dans les roches carbonatiques du Mésozoique des Carpates Occidentales par les courbes de tarissement des sources. Geologický ústav Dionýza Štúra, Bratislava. Západné Karpaty, sér. hydrogeológia a inžinierska geológia (3):7–60

Kullman E (1983) Režim podzemných vôd s turbulentným prúdením v puklinovo–krasovom horninovom prostredí [Groundwater regime with turbulent flow in fissure–karst rock media; in Slovak]. Geologický ústav Dionýza Štúra, Bratislava. Geologické práce (79):237–262

Kullman E (1990) Krasovo–puklinové vody [Karst-fissure waters]. Geologický ústav Dionýza Štúra, Bratislava, pp 184

Kullman E (2000) Nové metodické prístupy k riešeniu ochrany a ochranných pásiem zdrojov podzemných vôd v horninových prostrediach s krasovo-puklinovou a puklinovou priepustnosťou [New methodics for solution of protection and protection zones of the groundwater sources in rock environment with karst-joint and joint permeabilities, in Slovak]. Podzemná voda 6(2):31–41

Lamb R, Beven K (1997) Using interactive recession curve analysis to specify a general catchment storage model. Hydrol Earth Syst Sci 1(1):101–113

Maillet E (1905) Essais d'hydraulique souteeraine et fluviale. Hermann, Paris, p 218

Malík P (2007) Assessment of regional karstification degree and groundwater sensitivity to pollution using hydrograph analysis in the Velka Fatra Mts., Slovakia. Water resources and environmental problems in Karst. Environ Geol 51:707–711

Malík P (2010) Separácia hydrogramu pomocou parametrov reprezentatívnej výtokovej čiary [Hydrograph separation into flow components using parameters of the master recession curve, in Slovak with English summary]. Podzemná voda 16(1):113–124

Padilla A, Pulido Bosch A, Mangin A (1994) Relative importance of baseflow and quickflow from hydrographs of karst spring. Ground Water 32:267–277

Posavec K, Bačani A, Nakić Z (2006) A visual basic spreadsheet macro for recession curve analysis. Ground Water 44(5):764–767

Rutledge AT (1998) Computer Programs for describing the recession of ground-water discharge and for estimating mean ground-water recharge and discharge from streamflow records: update. Water-resources investigation report 98-4148, U.S. Geological Survey, pp 43

Samuel MP, Jha M (2003) Estimation of aquifer parameters from pumping test data by genetic algorithm optimization technique. J Irrig Drain Eng 129(5):348–359

Sharma V, Swayne DA, Lam D, Schertzer W (2006) Parallel shuffed complex evolution algortihm for calibration of hydrological models. In: The proceedings of the 20th international symposium on high-performance computing in an advanced collaborative environment (HPCS'06), Newfoundland, pp 30–35

Schoeller H (1965) Hydrodynamique dans le Karst. Hydrogeologie des roches fissurées, Actes du colloque de Dubrovnik 1965, IAHS-UNESCO

Tallaksen LM, van Lanen HAJ (eds) (2004) Hydrological drought, processes and estimation methods for streamflow and groundwater. Developments in water science, vol 48. Elsevier Science B.V, Amsterdam, pp 579

Well Hydrograph Analysis for the Estimation of Hydraulic and Geometric Parameters of Karst and Connected Water Systems

Attila Kovács and Pierre Perrochet

Abstract Well hydrographs contain crucial information about the hydraulic parameters and geometric characteristics of karst aquifers and connected water systems. This chapter provides quantitative tools for the estimation of hydraulic and geometric parameters by means of well hydrograph analysis. The analytical formulae provided in this chapter establish links between aquifer properties and hydrograph recession coefficients, and describe the spatial and temporal variations of the water table. A first set of equations describe the recession limb of hydrograph peaks, while another formula provides a quantitative characterization of the entire hydrograph peaks as a response to diffuse recharge. While existing spring hydrograph analytical techniques provide information on the overall characteristics of a karst catchment, well hydrograph analysis provides information on the hydraulic and geometric characteristics of individual matrix blocks. The combination of the spring and well hydrograph analytical techniques provides a powerful tool for the characterization of the structure and hydraulic behaviour of karst and connected water systems. A new approach to well hydrograph decomposition is presented, which makes the estimation of exact block geometry possible. In most cases, well hydrograph peaks can be decomposed into three exponential segments. These segments, however, do not represent different types of storage as suggested by previous studies. The asymmetric analytical solution presented in this study represents a powerful tool for parameter estimation in both karst and connected water systems. The symmetric and 1D solutions can be applied for order of magnitude parameter estimation. The proposed investigation method provides useful information for water resource assessment, flood prediction, vulnerability assessment, contamination risk assessment, geotechnical and speleological studies.

A. Kovács (✉)
Geological and Geophysical Institute of Hungary, 14 Stefania ut, Budapest 1143, Hungary
e-mail: attila.kovacs@unine.ch

P. Perrochet
CHYN, University of Neuchâtel, Rue Emile-Argand 11, 2007 Neuchâtel, Switzerland
e-mail: Pierre.Perrochet@unine.ch

J. Mudry et al. (eds.), *H2Karst Research in Limestone Hydrogeology*,
Environmental Earth Sciences, DOI: 10.1007/978-3-319-06139-9_7,

Keywords Karst • Well hydrograph • Hydrograph analysis • Conduit network • Parameter estimation • Connected water systems

1 Introduction

Karst aquifers are complex hydrogeological systems with strong heterogeneity, which originates from the presence of hierarchically organized dissolution channels. Heterogeneity manifests in the duality (diffuse and concentrated nature) of hydraulic processes taking place in karst (Király 1994) including recharge, groundwater flow and discharge.

The quantitative characterization of the hydrodynamic functioning of karst systems requires the definition of realistic hydraulic and geometric parameters (Király and Morel 1976; Király 1998, 2002). Direct information on these parameters is rarely available or difficult to obtain. It was estimated based on model calibration by Kovács (2003) that the explored sections of conduit systems do not represent more than a few percent of the total length of a conduit network, even in extensively studied cave systems. Also, there is usually very limited information available on hydraulically active conduits. Classical geological and hydrogeological surveys, borehole tests, tracing experiments, speleological and geophysical observations provide only limited information on the spatial configuration and hydraulic properties of a conduit network. However, in most cases well hydrograph data, coupled with information on the hydraulic properties of the low-permeability rock matrix, is available or easily obtainable via the installation of observation bores and well testing. Hydrograph analytical techniques presented here are suitable for estimating effective hydraulic parameters and geometric characteristics of karst and connected water systems.

The primary goal of the present chapter is to outline the theoretical bases of a quantitative method for the estimation of hydraulic parameters and conduit network geometry of karst aquifers and connected water systems based on well hydrograph analysis. This chapter does not aim at discussing field applications; it will be done in a subsequent publication.

2 Precedents

Spring hydrograph analytical methods for parameter estimation were provided by several authors (Rorabaugh 1964; Berkaloff 1967; Bagaric 1978; Kovács 2003; Kovács et al. 2005). Spring hydrograph analytical methods for the estimation of conduit network geometry were discussed by Kovács (2003), Kovács et al. (2005) and Kovács and Perrochet (2008).

Well hydrograph analysis for the estimation of hydraulic properties in karst was applied by Rorabough (1960), Atkinson (1977), Shevenell (1996), Powers and Shevenell (2000). These parameter estimation studies were mainly focussed on practical applications, and provided little theoretical background to well hydrograph behaviour. No attempt has been made so far to estimate block geometry based on well hydrograph data.

It was assumed by Shevenell (1996) that exponential components of well hydrographs represent three types of storage upgradient of the monitored point, similarly to the spring hydrograph decomposition theory of Forkasiewicz and Paloc (1967). This assumption will be further investigated in this study.

3 Discussion

The notion of the recession coefficient was first introduced by Maillet (1905). Maillet's interpretation is based on the drainage of a prismatic reservoir, and presumes that spring discharge is a linear function of the volume of water held in storage:

$$Q = \alpha V \tag{1}$$

where Q is discharge [L^3T^{-1}], V is volume of water in the reservoir [L^3], α is recession coefficient [T^{-1}] usually expressed in days. Under recession conditions, discharge also corresponds to the decrease of storage water:

$$Q = -\frac{dV}{dt} \tag{2}$$

Equating (1) and (2) and integrating during the recession time yield the Maillet formula:

$$Q_{(t)} = Q_0 e^{-\alpha t} \tag{3}$$

where Q_0 is the flow rate at the start of the recession period.

Assuming that the volume of water held in storage is a linear function of the reservoir area A and the water level H prevailing in it (i.e. V = AH), it can easily be shown that this level also follows an exponential behaviour with the same recession coefficient, namely:

$$H_{(t)} = H_0 e^{-\alpha t} \tag{4}$$

where H_0 is the head at the start of the recession period.

3.1 Conceptual Model

It is demonstrated above that recession coefficient can equally be expressed from discharge and hydraulic head time series for a simple prismatic reservoir with unit storage. However, further evidence is required to demonstrate the hydraulic behaviour of a strongly heterogeneous system. This is done by the application of a simple conceptual model of karst and connected water systems consisting of a rectangular catchment, a regular network of high-conductivity conduits embedded in the low-permeability rock matrix, and a single outlet (spring) which drains the entire conduit network. This model can be characterized by the hydraulic

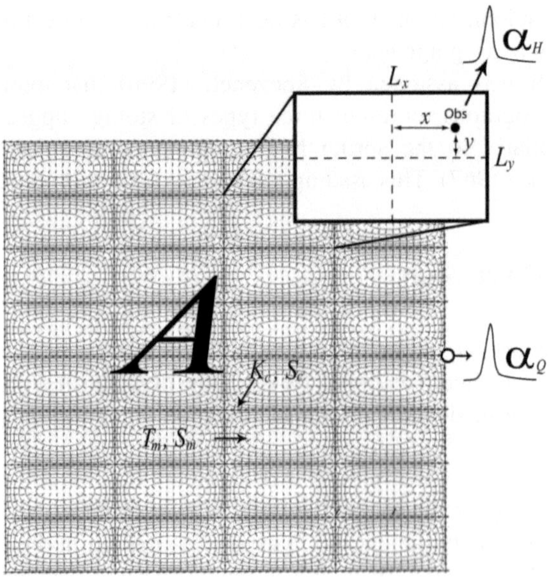

Fig. 1 A simple conceptual model suitable for the quantitative characterization of karst and connected hydrogeological systems. $T_m[L^2T^{-1}]$ transmissivity of the low-permeability matrix, $S_m[-]$ storativity of the low-permeability matrix, $K_c[L^3T^{-1}]$ 1D conduit conductance, $S_c[L]$ 1D conduit storativity, $A[L^2]$ spatial extent of the aquifer, L_x and $L_y[L]$ block size, x and y distance of piezometer from block centre, α_Q spring hydrograph recession coefficient, α_H well hydrograph recession coefficient

parameters of the low-permeability matrix and of the conduit system, conduit spacing, and the spatial extent of the water catchment (Fig. 1). This model represents an improved version of the conceptual model of Kovács et al. (2005).

3.2 Model Assumptions

As the applied conceptual model was developed for the purpose of deriving analytical solutions, it involves a number of simplifications:

- Two-dimensional flow assumption is applied: the provided two-dimensional solutions are generally applicable to shallow karst systems, where a single-level active conduit network exists generally at the base of the aquifer represented by an impervious layer or at the water table. Further considerations are necessary when the presented formulae are applied to multi-level or deep phreatic conduit systems.
- Karst conduits are considered as fixed-head boundary conditions: the fluctuations of water level in the karst conduit network are considered to be negligible compared to the changes of the water table in matrix blocks throughout the recession process. As a consequence, the precision of the parameter estimation method might decrease in the direct vicinity of karst conduits or surface streams, where flood recharge can reduce or reverse hydraulic gradients (Király and Morel 1976) during early times of the recession process.
- Uniform initial head conditions are applied: some of the analytical solutions provided in this chapter are based on uniform head initial conditions. This might result in inaccuracies at the very early stage of the recession process when applied to natural systems.

- Conduit blocks are rectangular: as karst conduit networks are usually pre-determined by conjugated networks of tectonic discontinuities, rectangular block shape is considered to be a satisfactory approximation. In the case of fractures/conduits intersecting at sharp angles, block shape can be represented by an asymmetric rectangular block formulation, for the calculation of fracture spacing.
- Uniform recharge function is applied: although the application of uniform recharge (both in time and space) at the catchment scale might be simplistic, it is a reasonable approximation at the block scale. The temporal variation of recharge primarily influences the rising limb of a hydrograph peak. As the presented parameter estimation method is based on falling limb hydrograph segments, the above assumption has limited implications on the applicability of this method.

In order to test the applicability of the analytical formulae developed, numerical simulations of synthetic karst systems were performed and results were compared with those of the analytical calculations. The advantage of this method as opposed to using field data is that the internal structure and hydraulic properties of our numerical system are known, while these are largely unknown in case of real systems.

3.3 Criteria for Karstic Hydraulic Behaviour

Although there is a fundamental difference in the formation processes between karst and fractured hydrogeological systems, their hydraulic behaviour cannot be determined purely based on their geological characteristics. This is especially valid for dolomitic rock formations, which are often considered as fractured systems, treated as porous media and called as karst aquifers. The transient hydraulic behaviour of strongly heterogeneous (karst or fractured) systems depends on their degree of heterogeneity which can be quantified by a combination of hydraulic and geometric properties as demonstrated by Kovács (2003) and Kovács et al. (2005).

The baseflow discharge of well-developed karst systems is controlled by the drainage of individual matrix blocks. This flow condition has been referred to as matrix-restrained flow regime (MRFR).

The baseflow discharge of fractured systems is influenced by the hydraulic parameters of fractures/conduits, low-permeability blocks, fracture spacing, and aquifer extent. This flow condition has been defined as conduit-influenced flow regime (CIFR).

A threshold value for conduit conductance between these two domains can be expressed as follows (Fig. 2):

$$K_c^* \approx 3\pi^2 T_m A f \tag{5}$$

where $T_m[L^2T^{-1}]$ is matrix transmissivity, $K_c[L^3T^{-1}]$ is 1D conduit conductance, $A[L^2]$ is aquifer extent, $f[L^{-1}]$ is average conduit frequency.

The analytical formulae presented in this chapter are only valid for karst systems (MRFR regime) where $K_c > K_c^*$. These following observations are not valid for fractured systems (CIFR regime).

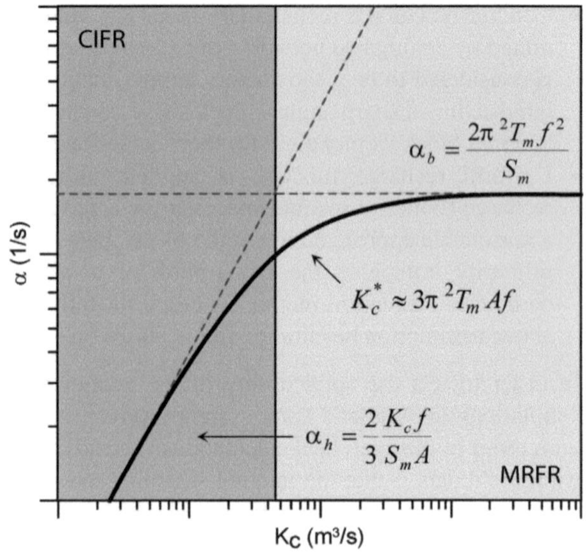

Fig. 2 Dependence of the baseflow recession coefficient on conduit conductance. K_c* is the threshold value between MRFR and CIFR baseflow. From Kovács et al. (2005)

4 Hydraulic Head Distribution in Rectangular Matrix Blocks

The distribution of hydraulic heads in the matrix blocks of a karst or connected water system is strongly time dependent; primarily influenced by initial heads and the duration, intensity and temporal variations of diffuse infiltration over the aquifer surface. A first set of analytical solutions provided here were developed to describe the evolution of hydraulic heads over the low permeability matrix blocks during a recession period. This assumption included uniform-head initial conditions across a matrix block. A second type of analytical solution provides a quantitative characterization of the hydraulic potential field during and following a constant recharge event. A comprehensive numerical analysis was undertaken to test the validity and applicability of these solutions in synthetic karst aquifer configurations.

4.1 Analytical Solution: Uniform Initial Head Distribution Over an Asymmetric Block

In order to provide a mathematical characterization of hydraulic head distribution over rectangular matrix blocks of karst and connected water systems, the following 2D analytical solution has been derived:

$$H_{(x,y,t)} = \frac{16H_0}{\pi^2}\sum_{m=0}^{\infty}\frac{(-1)^m\cos((2m+1)\frac{\pi}{L_x}x)}{(2m+1)}e^{-a(2m+1)^2}$$
$$\sum_{n=0}^{\infty}\frac{(-1)^n\cos\left((2n+1)\frac{\pi}{L_y}y\right)}{(2n+1)}e^{-a\beta^2(2n+1)^2} \quad (6)$$

where:

$$\alpha = \frac{\pi^2 Tt}{SL_x^2} \quad \text{and} \quad \beta = \frac{L_x}{L_y}$$

Assuming uniform initial condition H_0 over the block surface, as well as zero hydraulic head as boundary conditions along the sides of the blocks representing karst conduits or connected streams, this solution is written down as the product, in the x and y directions, of the 1D solution given in Carslaw and Jaeger (1959, p. 144 and 173).

It follows from Eq. (6) that:

$$\begin{aligned} H_{(x,y,t)} = \frac{16H_0}{\pi^2}&\left(\cos\left(\frac{\pi x}{L_x}\right)e^{-a} - \frac{1}{3}\cos\left(\frac{3\pi x}{L_x}\right)e^{-9a}\right. \\ &\left.+\frac{1}{5}\cos\left(\frac{5\pi x}{L_x}\right)e^{-25a}-\ldots\right) \\ &\left(\cos\left(\frac{\pi y}{L_y}\right)e^{-a\beta^2} - \frac{1}{3}\cos\left(\frac{3\pi y}{L_y}\right)e^{-9a\beta^2}\right. \\ &\left.+\frac{1}{5}\cos\left(\frac{5\pi y}{L_y}\right)e^{-25a\beta^2}-\ldots\right) \end{aligned} \quad (7)$$

Leading to the expansion:

$$H_{(x,y,t)} = \frac{16H_0}{\pi^2}\left(\begin{array}{l} \cos\left(\frac{\pi x}{L_x}\right)\cos\left(\frac{\pi y}{L_y}\right)e^{-a(1+\beta^2)} - \frac{1}{3}\cos\left(\frac{\pi x}{L_x}\right)\cos\left(\frac{3\pi y}{L_y}\right)e^{-a(1+9\beta^2)} \\ +\frac{1}{5}\cos\left(\frac{\pi x}{L_x}\right)\cos\left(\frac{5\pi y}{L_y}\right)e^{-a(1+25\beta^2)} \\ -\frac{1}{3}\cos\left(\frac{3\pi x}{L_x}\right)\cos\left(\frac{\pi y}{L_y}\right)e^{-a(9+\beta^2)} + \frac{1}{9}\cos\left(\frac{3\pi x}{L_x}\right)\cos\left(\frac{3\pi y}{L_y}\right)e^{-9a(1+\beta^2)} \\ -\frac{1}{15}\cos\left(\frac{3\pi x}{L_x}\right)\cos\left(\frac{5\pi y}{L_y}\right)e^{-a(9+25\beta^2)} + \frac{1}{5}\cos\left(\frac{5\pi x}{L_x}\right)\cos\left(\frac{\pi y}{L_y}\right)e^{-a(25+\beta^2)} \\ -\frac{1}{15}\cos\left(\frac{5\pi x}{L_x}\right)\cos\left(\frac{3\pi y}{L_y}\right)e^{-a(25+9\beta^2)} + \frac{1}{25}\cos\left(\frac{5\pi x}{L_x}\right)\cos\left(\frac{5\pi y}{L_y}\right)e^{-25a(1+\beta^2)}\ldots \end{array}\right)$$

(8)

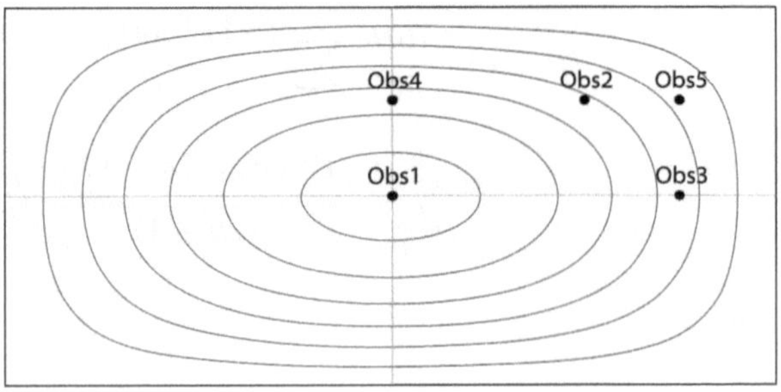

Fig. 3 Position of observation points in numerical block model. $L_x = 1{,}200$, $L_y = 600$ m

Table 1 Distances of observation points from the block centre in numerical block model

Observation point	X	y
Obs1	0	0
Obs2	300	150
Obs3	450	0
Obs4	0	150
Obs5	450	150

$L_x = 1{,}200$, $L_y = 600$ m

Analytical hydrographs were compared with numerically simulated hydrographs for five different observation points in an asymmetric block. Block properties included $T = 1e{-}5$ m^2/s, $S = 1e{-}4$, $H_0 = 100$ m, $L_x = 1{,}200$, $L_y = 600$ m. The positions of observation points within the asymmetric block are indicated in Fig. 3 and Table 1. Figure 4 indicates a very good agreement between numerical and analytical results, and verifies the validity of the analytical formula (Eq. 11).

Figure 5 indicates the first nine hydrograph components ($n = m = 3$ in Eq. 6) for observation point Obs5. For any asymmetric block, the first component is H1, the second component is H2 (for $L_y > Lx$) or H4 (for $L_y < Lx$), where H2 = H4 in the block centre of symmetric blocks ($Lx = Ly$ and $x = y = 0$). The third component is H3 (for $L_y > Lx$) or H7 (for $L_y < Lx$). As H2 and H4 can equally take negative or positive values depending on x and y, a modified approach to hydrograph decomposition can be applied for determining the asymmetry factor (β) from the first two components of well hydrographs.

If the second exponential component (H2 or H4) is positive (concave hydrograph peak), the traditional decomposition technique of Forkasiewicz and Paloc (1967) originally introduced for spring hydrographs can be applied. In the case if the second exponential component (H2 or H4) is negative (convex hydrograph peak), a complementer decomposition approach can be applied. After fitting a straight line on the semilog graph of the “baseflow” well hydrograph, the remaining hydrograph needs to be subtracted from the first exponential. A second exponential can then be determined by line fitting on the residual hydrograph. In most cases a third exponential can be fitted, which can also be used for parameter estimation purposes. The asymmetry factor can be estimated through hydrograph decomposition both in the convex and concave cases from the first two exponentials as follows:

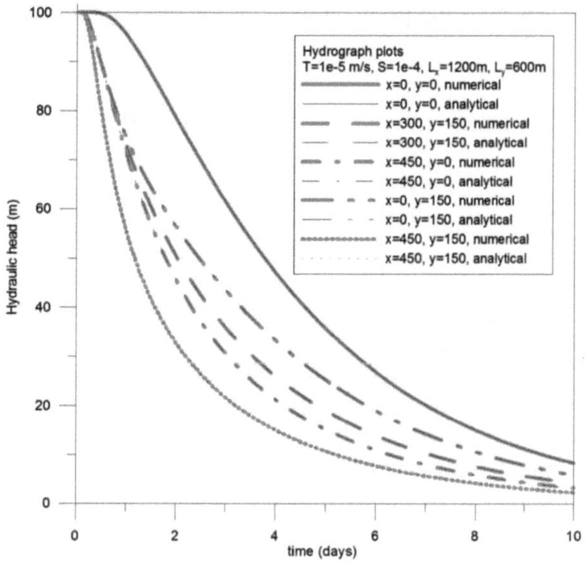

Fig. 4 Analytical versus numerical well hydrographs (T = 1e−5 m^2/s, S = 1e−4, H_0 = 100 m, L_x = 1,200, L_y = 600 m)

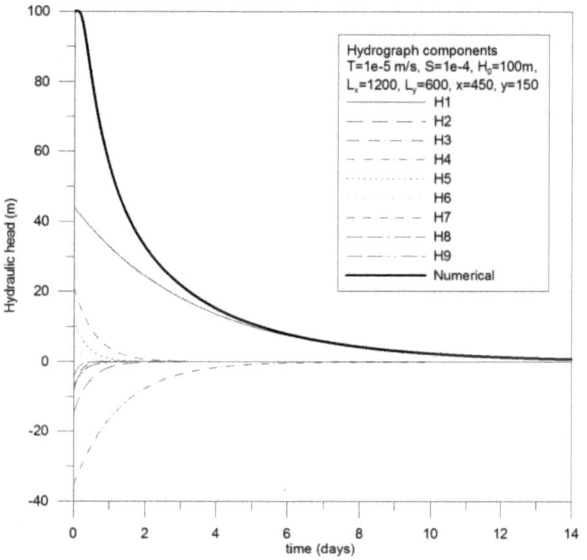

Fig. 5 Analytical well hydrographs components for Obs5 (T = 1e−5 m^2/s, S = 1e−4, H_0 = 100 m, L_x = 1,200, L_y = 600 m, x = 450, y = 150)

$$\beta = \sqrt{\frac{\gamma - 1}{1 - 9\gamma}} \quad \text{if } L_y > L_x$$

$$\beta = \sqrt{\frac{9\gamma - 1}{1 - \gamma}} \quad \text{if } L_x > L_y \tag{9}$$

Or from the first and third exponentials as follows:

$$\beta = \sqrt{\frac{\delta - 1}{1 - 25\delta}} \quad \text{if } L_y > L_x$$

$$\beta = \sqrt{\frac{25\delta - 1}{1 - \delta}} \quad \text{if } L_x > L_y \tag{10}$$

where $\gamma = \frac{\alpha_1}{\alpha_2}$, $\delta = \frac{\alpha_1}{\alpha_3}$, α_1, α_2 and α_3 are recession coefficients of the first, second and third exponential components.

The asymmetry factor determined from a well hydrograph provides information on the block shape of an individual matrix block. The asymmetry factor determined from a spring hydrograph (Kovács and Perrochet 2008) provides an estimation of average block asymmetry throughout a catchment area, and might be different from the asymmetry factor of individual matrix blocks. For a sufficient dataset, a distribution function of block asymmetry can be determined, which can serve as a means for a stochastic generation of conduit networks.

If block asymmetry is estimated via well hydrograph decomposition, and hydraulic tests can be performed in the same or nearby wells, a quantitative estimation of conduit spacing (or block size) can be made from one of the recession coefficients. Based on Eq. 8 the recession coefficient of the first exponential component can be expressed as follows:

$$\alpha_1 = \frac{\pi^2 T}{S L_x^2}\left(1 + \beta^2\right) = \frac{\pi^2 T}{S}\left(\frac{1}{L_x^2} + \frac{1}{L_y^2}\right) \tag{11}$$

And the second exponential can be expressed as:

$$\alpha_2 = \frac{\pi^2 T}{S L_x^2}\left(1 + 9\beta^2\right) = \frac{\pi^2 T}{S}\left(\frac{1}{L_x^2} + \frac{9}{L_y^2}\right) \quad \text{for } L_y > L_x \tag{12}$$

or

$$\alpha_2 = \frac{\pi^2 T}{S L_x^2}\left(9 + \beta^2\right) = \frac{\pi^2 T}{S}\left(\frac{9}{L_x^2} + \frac{1}{L_y^2}\right) \quad \text{for } L_x > L_y \tag{13}$$

The third exponential can be expressed as:

$$\alpha_3 = \frac{\pi^2 T}{S L_x^2}\left(1 + 25\beta^2\right) = \frac{\pi^2 T}{S}\left(\frac{1}{L_x^2} + \frac{25}{L_y^2}\right) \quad \text{for } L_y > L_x$$

Or

$$\alpha_3 = \frac{\pi^2 T}{S L_x^2}\left(25 + \beta^2\right) = \frac{\pi^2 T}{S}\left(\frac{25}{L_x^2} + \frac{1}{L_y^2}\right) \quad \text{for } L_x > L_y \tag{14}$$

The same formulae can be applied for estimating the hydraulic parameters of connected water systems based on the spacing of surface water streams and well hydrograph data.

4.2 Analytical Solution : Uniform Initial Head Distribution Over a Symmetric Block

A symmetric approximation of block shape (square blocks) can be applied in the case of negligible anisotropy with quasi-symmetric matrix blocks. This solution is also useful when only the "baseflow" recession coefficient (α_1) can be estimated because of insufficient or bad quality data. In such cases block asymmetry (β) cannot be calculated, and thus the application of the asymmetric solution would be problematic. The following symmetric solution provides an efficient tool for parameter estimation is these situations. For symmetric blocks, Eq. 6 takes the following form (n = m = 2):

$$H = \frac{16H_0}{\pi^2}\left(\begin{array}{l} \cos\left(\frac{\pi x}{L}\right)\cos\left(\frac{\pi y}{L}\right)e^{-2a} - \frac{1}{3}\left[\cos\left(\frac{\pi x}{L}\right)\cos\left(\frac{3\pi y}{L}\right)\right. \\ \left.+\cos\left(\frac{3\pi x}{L}\right)\cos\left(\frac{\pi y}{L}\right)\right]e^{-10a} + \frac{1}{5}\left[\cos\left(\frac{\pi x}{L}\right)\cos\left(\frac{5\pi y}{L}\right)\right. \\ \left.+\cos\left(\frac{5\pi x}{L}\right)\cos\left(\frac{\pi y}{L}\right)\right]e^{-26a} + \frac{1}{9}\cos\left(\frac{3\pi x}{L}\right)\cos\left(\frac{3\pi y}{L}\right)e^{-18a} \\ -\frac{1}{15}\left[\cos\left(\frac{3\pi x}{L}\right)\cos\left(\frac{5\pi y}{L}\right) + \cos\left(\frac{5\pi x}{L}\right)\cos\left(\frac{3\pi y}{L}\right)\right]e^{-34a} \\ +\frac{1}{25}\cos\left(\frac{5\pi x}{L}\right)\cos\left(\frac{5\pi y}{L}\right)e^{-50a} - \ldots \end{array}\right) \tag{15}$$

It follows from Eq. 15 that "baseflow" recession coefficient can be expressed from the first exponential component as:

$$\alpha_1 = \frac{2\pi^2 T}{SL^2} \tag{16}$$

While subsequent exponentials can be expressed as:

$$\alpha_2 = \frac{10\pi^2 T}{SL^2} \tag{17}$$

$$\alpha_3 = \frac{18\pi^2 T}{SL^2} \tag{18}$$

Consequently, the "baseflow" sections of well hydrographs can be used for estimating block size (if block hydraulic parameters are known), or hydraulic diffusivity of matrix blocks (if block size is known) of a karst or connected water system. Most importantly, hydrograph analysis represents an additional tool for verifying assumptions on the geometric and hydraulic parameters of complex hydrogeological systems.

According to Eqs. 16–18, similar formulae can be used for expressing recession coefficients from well hydrographs as from spring hydrographs (Kovács et al. 2005; Kovács and Perrochet 2008). However, the decomposition method is different for the two data types, and while the former represents block properties, the latter one provides information on the overall characteristics of an entire karstic catchment.

4.3 Analytical Solution: Uniform Initial Head Distribution Over a 1D Block

A 1D approximation of block shape can theoretically be applied in the case of extreme anisotropy with elongated matrix blocks. According to this solution, it is assumed that the influence of the long side of a matrix block has negligible influence on the hydraulic head distribution within the block at the observation point. Hydraulic head as a function of time and distance from block centre can be expressed as follows:

$$H_{(x,t)} = \frac{4H_0}{\pi} \sum_{n=0}^{\infty} \frac{(-1)^n \cos\left((2n+1)\frac{\pi}{2L}x\right)}{(2n+1)} e^{-\frac{a}{4}(2n+1)^2} \tag{19}$$

where

$$a = \frac{\pi^2 Tt}{SL^2} \tag{20}$$

L represents the half of the short side of a strongly asymmetric block. For n = 3:

$$\begin{aligned} H_{(x,t)} = \frac{4H_0}{\pi} \Bigg(&\cos\left(\frac{1}{2}\frac{\pi x}{L}\right) e^{-\frac{1}{4}a} - \frac{1}{3}\cos\left(\frac{3}{2}\frac{\pi x}{L}\right) e^{-\frac{9}{4}a} \\ &+ \frac{1}{5}\cos\left(\frac{5}{2}\frac{\pi x}{L}\right) e^{-\frac{25}{4}a} - \frac{1}{7}\cos\left(\frac{7}{2}\frac{\pi x}{L}\right) e^{-\frac{49}{4}a} + \cdots \Bigg) \end{aligned} \tag{21}$$

It follows from Eq. 15 that “baseflow” recession coefficient can be expressed from the first exponential component as:

$$\alpha_1 = \frac{\pi^2 T}{4SL^2} \tag{22}$$

While subsequent exponentials can be expressed as:

$$\alpha_2 = \frac{9\pi^2 T}{4SL^2} \tag{23}$$

$$\alpha_3 = \frac{25\pi^2 T}{4SL^2} \tag{24}$$

4.4 Analytical Solution: Uniform Recharge Over a 2D Asymmetric Block

An analytical solution for the quantitative characterization of the hydraulic potential field over an asymmetric rectangular block during and following a constant recharge event has been derived using the elementary constant recharge solution developed by Bruggeman (1999, p. 323). The principle of superposition has been applied in time to account for a recharge event of finite duration τ as follows:

$$H_{(x,y,t)} = \frac{16 i_0 L_x^2}{\pi^4 T} \sum_{m=0}^{\infty} \sum_{n=0}^{\infty} \frac{(-1)^{m+n} \cos\left((2m+1)\frac{\pi}{L_x}x\right) \cos((2n+1)\frac{\pi}{L_y}y)}{((2m+1)^2 + \beta^2(2n+1)^2)(2m+1)(2n+1)} \begin{aligned} &(1 - e^{-a((2m+1)^2+\beta^2(2n+1)^2)} \\ &- H(a-a_\tau)(1 - e^{-(a-a_\tau)((2m+1)^2+\beta^2(2n+1)^2)})) \end{aligned} \tag{25}$$

where

$$a_\tau = \frac{\pi^2 T \tau}{S L_x^2}$$

where τ is the time corresponding to the end of the constant recharge, i_0 is the recharge rate (LT^{-1}) and $H(a-a_\tau)$ is the Heaviside function that takes the value $H = 0$ for $a \leq a_\tau$ and $H = a - a_\tau$ for $a > a_\tau$. From Eq. (25) if follows that the rising limb hydraulic head is:

$$H_{(x,y,t)} = \frac{16 i_0 L_x^2}{\pi^4 T} \left(\begin{aligned} &\cos\left(\frac{\pi x}{L_x}\right)\cos\left(\frac{\pi y}{L_y}\right)\frac{(1-e^{-a(1+\beta^2)})}{(1+\beta^2)} \\ &-\frac{1}{3}\cos\left(\frac{3\pi x}{L_x}\right)\cos\left(\frac{\pi y}{L_y}\right)\frac{(1-e^{-a(9+\beta^2)}}{(9+\beta^2)}) \\ &+\frac{1}{5}\cos\left(\frac{5\pi x}{L_x}\right)\cos\left(\frac{\pi y}{L_y}\right)\frac{(1-e^{-a(25+\beta^2)})}{(25+\beta^2)} \\ &-\frac{1}{3}\cos\left(\frac{\pi x}{L_x}\right)\cos\left(\frac{3\pi y}{L_y}\right)\frac{(1-e^{-a(1+9\beta^2)})}{(1+9\beta^2)} \\ &+\frac{1}{9}\cos\left(\frac{3\pi x}{L_x}\right)\cos\left(\frac{3\pi y}{L_y}\right)\frac{(1-e^{-a(9+9\beta^2)})}{(9+9\beta^2)} \\ &-\frac{1}{15}\cos\left(\frac{5\pi x}{L_x}\right)\cos\left(\frac{3\pi y}{L_y}\right)\frac{(1-e^{-a(25+9\beta^2)})}{(25+9\beta^2)} \\ &+\frac{1}{5}\cos\left(\frac{\pi x}{L_x}\right)\cos\left(\frac{5\pi y}{L_y}\right)\frac{(1-e^{-a(1+25\beta^2)})}{(1+25\beta^2)} \\ &-\frac{1}{15}\cos\left(\frac{3\pi x}{L_x}\right)\cos\left(\frac{5\pi y}{L_y}\right)\frac{(1-e^{-a(9+25\beta^2)})}{(9+25\beta^2)} \\ &+\frac{1}{25}\cos\left(\frac{5\pi x}{L_x}\right)\cos\left(\frac{5\pi y}{L_y}\right)\frac{(1-e^{-a(25+25\beta^2)})}{(25+25\beta^2)} \cdots \end{aligned} \right) \tag{26}$$

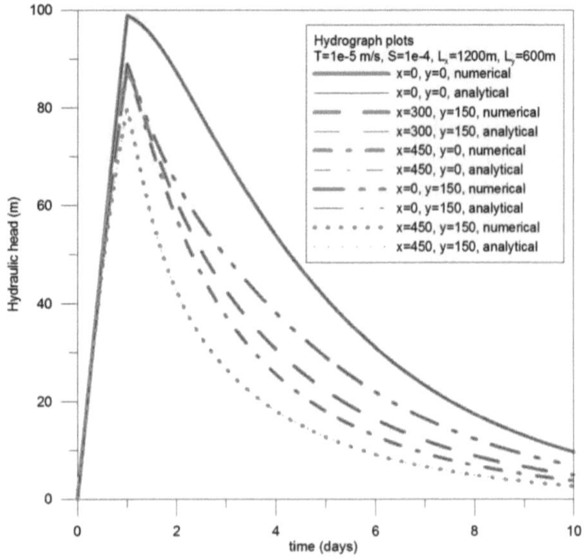

Fig. 6 Analytical versus numerical well hydrographs (T = 1e−5 m²/s, S = 1e−4, i_0 = 1.16e−7 m/s, H_0 = 0 m, L_x = 1,200, L_y = 600 m)

And the falling limb head is:

$$H_{(x,y,t)} = \frac{16 i_0 L_x^2}{\pi^4 T} \left(\begin{array}{l} \cos\left(\frac{\pi x}{L_x}\right) \cos\left(\frac{\pi y}{L_y}\right) \frac{(-e^{-a(1+\beta^2)} + e^{-(a-a_\tau)(1+\beta^2)})}{(1+\beta^2)} \\ -\frac{1}{3} \cos\left(\frac{3\pi x}{L_x}\right) \cos\left(\frac{\pi y}{L_y}\right) \frac{(-e^{-a(9+\beta^2)} + e^{-(a-a_\tau)(9+\beta^2)})}{(9+\beta^2)}) \\ +\frac{1}{5} \cos\left(\frac{5\pi x}{L_x}\right) \cos\left(\frac{\pi y}{L_y}\right) \frac{(-e^{-a(25+\beta^2)} + e^{-(a-a_\tau)(25+\beta^2)})}{(25+\beta^2)} \\ -\frac{1}{3} \cos\left(\frac{\pi x}{L_x}\right) \cos\left(\frac{3\pi y}{L_y}\right) \frac{(-e^{-a(1+9\beta^2)} + e^{-(a-a_\tau)(1+9\beta^2)})}{(1+9\beta^2)} \\ +\frac{1}{9} \cos\left(\frac{3\pi x}{L_x}\right) \cos\left(\frac{3\pi y}{L_y}\right) \frac{(-e^{-a(9+9\beta^2)} + e^{-(a-a_\tau)(9+9\beta^2)})}{(9+9\beta^2)} \\ -\frac{1}{15} \cos\left(\frac{5\pi x}{L_x}\right) \cos\left(\frac{3\pi y}{L_y}\right) \frac{(-e^{-a(25+9\beta^2)} + e^{-(a-a_\tau)(25+9\beta^2)})}{(25+9\beta^2)} \\ +\frac{1}{5} \cos\left(\frac{\pi x}{L_x}\right) \cos\left(\frac{5\pi y}{L_y}\right) \frac{(-e^{-a(1+25\beta^2)} + e^{-(a-a_\tau)(1+25\beta^2)})}{(1+25\beta^2)} \\ -\frac{1}{15} \cos\left(\frac{3\pi x}{L_x}\right) \cos\left(\frac{5\pi y}{L_y}\right) \frac{(-e^{-a(9+25\beta^2)} + e^{-(a-a_\tau)(9+25\beta^2)})}{(9+25\beta^2)} \\ +\frac{1}{25} \cos\left(\frac{5\pi x}{L_x}\right) \cos\left(\frac{5\pi y}{L_y}\right) \frac{(-e^{-a(25+25\beta^2)} + e^{-(a-a_\tau)(25+25\beta^2)})}{(25+25\beta^2)} \ldots \end{array} \right) \quad (27)$$

A comprehensive numerical analysis was undertaken to test the validity and applicability of these solutions in synthetic karst aquifer configurations. Analytical and numerical solutions for five observation points in an asymmetric block are indicated in Fig. 3. The numerical analyses verified the validity of the analytical formula.

5 Well Hydrograph Decomposition: A Synthetic Example

Numerical modelling has been used to demonstrate the applicability of the above hydrograph analytical techniques through a synthetic example. The advantage of a numerical example is that the applied parameters are known, and thus the results of hydrograph analysis can directly be verified.

The synthetic karst system applied consists of a square shaped karstic catchment of 4,800 m × 4,800 m size (Fig. 1). The spacing of karst conduits is 1,200 m in the x direction and 600 m in the y direction. The applied matrix transmissivity is $T_m = 1e{-5}\ m^2/s$, conduit conductance is $K_c = 1e + 2\ m^3/s$, matrix storativity is $S_m = 1e{-4}$, conduit storativity is $S_c = 2.14e{-7}$ m. A 24-h infiltration pulse of 20 mm/day has been applied throughout the model area representing a total infiltration of 460,800 m^3. Half of this recharge (1.16e−7 m/s) was distributed as diffuse recharge over the aquifer surface, while half of the total recharge (3.97e−5 m^2/s) was distributed along the karst conduit network as direct recharge. A constant recharge function has been applied. In order to reproduce realistic initial conditions, the same recharge pulse was applied twice with a 9-day gap in between. A uniform hydraulic head distribution of $H = 0$ m was applied as initial condition before the first recharge event. Only the second hydrograph peak was considered during hydrograph analysis.

Numerical simulations were performed by the FEN family of discrete-continuum finite element codes (Király 1985).

The hydraulic heads at an observation point at model coordinates $X = 2{,}850$, $Y = 4{,}575$ were exported and analysed for demonstration purposes. The block coordinates (relative to block centre) of this observation point are $x = 150$, $y = 75$. The falling limb of the simulated hydrograph at Obs1 and its decomposition are indicated in Fig. 7.

According to Fig. 7, the hydrograph at Obs1 can be manually decomposed into three exponential segments. A convex hydrograph peak indicates that the second component is negative.

In order to check the applicability of the above analytical solutions, a numerically simulated hydrograph peak at Obs1 was manually decomposed and recession coefficients $\alpha_{1,2,3}$ were used for calculating block geometry and diffusivity ($D = T/S$). Block size was calculated via well hydrograph analysis based on the assumption that diffusivity is determined from pump tests in a matrix block. This scenario represents a karst system, where hydrograph data together with hydraulic

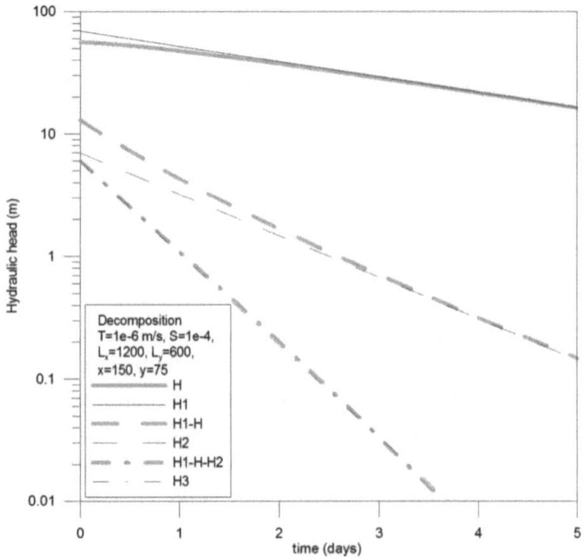

Fig. 7 Well hydrograph decomposition. $L_x = 1{,}200$ m, $L_y = 600$, $x = 150$, $y = 75$, $T_m = 1e{-}5\ m^2/s$, $S_m = 1e{-}4$

test data is collected over a matrix block, and block dimensions (conduit geometry) needs to be determined.

Diffusivity was calculated via well hydrograph analysis based on the assumption that block size is known. This scenario represents a connected water system, where the geometry of a river network around a connected aquifer block is known, and well hydrograph data is available from within this block.

Hydraulic and geometric parameters were calculated making use of the three analytical solutions presented in Sects. 1–3 to verify the applicability of these formulae and to demonstrate the accuracy of the well hydrograph analytical method.

Based on the comparative numerical simulations (Figs. 4 and 6) and the synthetic example (Fig. 7 and Table 2), the asymmetric solution provides an accurate tool for the estimation of block geometry and hydraulic properties. An average error of 2–4 % can be achieved, where the largest uncertainty originates from manual curve fitting performed during the decomposition process.

The symmetric and 1D solutions represent simple and quick tools for order of magnitude parameter estimation with an error believed to be inferior to 74 % (symmetric solution) and 91 % (1D solution) in the case of an asymmetry factor of $\beta = 2$. A much higher accuracy can be achieved when the symmetric solution is applied for quasi-symmetric systems. Similarly, accuracy improves if the 1D solution is used for strongly asymmetric blocks.

Where sufficient field data is available, the use of the asymmetric solution is recommended. The symmetric solution can be used where less information can be gathered. The 1D solution, which was used in previous studies, has limited applicability and must be applied with care.

Table 2 Hydraulic and geometric parameters obtained through well hydrograph decomposition using various analytical solutions

Parameter	Numerical	Asym solution	Error (%)	Sym solution	Error (%)	1D solution	Error (%)
α_1	3.33E−06	3.42E−06	3	2.43E−06	27	3.04E−07	91
α_2	9.00E−06	8.90E−06	1	1.22E−05	35	2.74E−06	70
α_3	2.00E−05	1.99E−05	1	2.19E−05	10	7.61E−06	62
β (from $\alpha_{1,2}$)	2.00E+00	1.92E+00	4	2.00E+00	N/A	N/A	N/A
β (from $\alpha_{1,3}$)	2.00E+00	1.95E+00	3	2.00E+00	N/A	N/A	N/A
D (from α_1)	1.00E−01	1.04E−01	4	1.37E−01	37	1.22E−01	22
D (from α_2)	1.00E−01	1.04E−01	4	7.39E−02	26	3.65E−02	63
D (from α_3)	1.00E−01	1.02E−01	2	9.13E−02	9	2.92E−02	71
Lx (from α_1)	1.20E+03	1.18E+03	2	7.70E+02	36	N/A	N/A
Lx (from α_2)	1.20E+03	1.18E+03	2	1.05E+03	13	N/A	N/A
Lx (from α_3)	1.20E+03	1.19E+03	1	9.42E+02	22	N/A	N/A
Ly (from α_1)	6.00E+02	6.13E+02	2	7.70E+02	28	5.44E+02	9
Ly (from α_2)	6.00E+02	6.13E+02	2	1.05E+03	74	9.93E+02	65
Ly (from α_3)	6.00E+02	6.18E+02	3	9.42E+02	57	1.11E+03	85

Recession coefficients were determined by means of manual curve fitting from a numerically simulated hydrograph ($T = 1e{-5}\ m^2/s$, $S = 1e{-4}$, $i_0 = 1.16e{-7}$ m/s, $L_x = 1{,}200$, $L_y = 600$ m, $x = 150$, $y = 75$)

6 Conclusions

The present chapter describes the theoretical basis of new quantitative methods for the estimation of hydraulic and geometric parameters of karst and connected water systems. It also provides a validation of these methods through comparison with numerical model results.

As spring hydrograph data are not always available, well hydrographs represent an alternative way to hydraulic parameter estimation. While spring hydrograph analytical methods assume a uniform parameter distribution and conduit spacing throughout a karstic catchment, well hydrographs facilitate the determination of the size and hydraulic parameters of matrix blocks at specific locations, and thus provide information on spatial variability. An additional advantage of the well hydrograph method compared to spring hydrographs is the negligible influence of concentrated infiltration on the dataset. Furthermore to this, the conduit locations determined from a sufficient number of well hydrographs make the estimation of conduit spacing distribution functions possible, which can be used for a stochastic generation of extensive conduit networks. The combination of the spring and well hydrograph analytical techniques provides a powerful tool for the characterization of the structure and hydraulic properties of karst and connected water systems.

Recession coefficient is an important indicator of block geometry and hydraulic characteristics. In most cases well hydrograph peaks can be decomposed into three exponential segments. These segments, however, do not necessarily represent different types of storage as suggested by previous studies. The recession coefficients determined from hydrograph decomposition serve for estimating block asymmetry and the specific dimensions of matrix blocks. The asymmetric analytical solution presented in this study represents a powerful tool for parameter estimation in both karst and connected water systems. The symmetric and 1D solutions can be applied for order of magnitude parameter estimation.

The proposed investigation method provides useful information for water resource assessment, flood prediction, vulnerability assessment, contamination risk assessment, geotechnical and speleological studies.

References

Atkinson TC (1977) Diffuse flow and conduit flow in limestone terrain in the Mendip Hills, Somerset (Great Britain). J Hydrol 35:93–103

Bagarić I (1978) Determination of storage and transportation characteristics of karst aquifers. In: Milanović PT (1981) Karst hydrogeology. Water Resources Publications, Littleton, 434 p

Berkaloff E (1967) Limite de validité des formules courantes de tarissement de débit. Chronique d'Hydrogéologie 10:31–41

Bruggeman GA (1999) Analytical solutions of geohydrological problems. Developments in water sciences, vol 46, Elsevier, Amsterdam, 959 p

Carslaw HS, Jaeger JC (1959) Conduction of heat in solids (2nd edn). Oxford University Press, London, 510 p

Forkasiewicz J, Paloc H (1967) Le régime de tarissement de la Foux-de-la-Vis. Etude préliminaire. Chronique d'Hydrogéologie, BRGM, 3(10):61–73

Király L (1985) FEM-301: a three-dimensional model for groundwater flow simulation. NAGRA technical report 84-49, 96 p

Király L (1994) Groundwater flow in fractures rocks: models and reality. In: 14th Mintrop Seminar über Interpretationsstrategien in Exploration und Produktion, Ruhr Universität Bochum 159:1–21

Király L (1998) Modeling karst aquifers by the combined discrete channel and continuum approach. Bulletin d'Hydrogéologie de l'Université de Neuchâtel 16:77–98

Király L (2002) Karstification and groundwater flow. In: Proceedings of the conference on evolution of karst: from Prekarst to Cessation. Postojna-Ljubljana, pp 155–190

Király L, Morel G (1976) Remarques sur l'hydrogramme des sources karstiques simule par modèles mathématiques. Bulletin d'Hydrogéologie de l'Université de Neuchâtel 1:37–60

Kovács A (2003) Geometry and hydraulic parameters of karst aquifers: a hydrodynamic modeling approach. Doctoral thesis, University of Neuchâtel, Switzerland, 131 p

Kovács A, Perrochet P (2008) A quantitative approach to spring hydrograph decomposition. J Hydrol 352:16–29

Kovács A, Perrochet P, Király L, Jeannin P-Y (2005) A quantitative method for the characterization of karst aquifers based on spring hydrograph analysis. J Hydrol 303:152–164

Maillet E (1905) Essais d`hydraulique souterraine et fluviale. Hermann, Paris

Powers JG, Shevenell L (2000) Evaluating transmissivity estimates from well hydrographs in karst aquifers. Ground Water 38(3):361–369

Rorabaugh MI (1964) Changes in bank storage. Publication No. 63, IASH, Gentbrugge, pp 432–441

Rorabough MI (1960) Use of water levels in estimating aquifer constants in a finite aquifer. Int Assoc Sci Hydrol 52:314–323

Shevenell L (1996) Analysis of well hydrographs in a karst aquifer: estimates of specific yields and continuum transmissivities. J Hydrol 174:331–355

Distributed Hydrological Modeling and Model Adaption in High Alpine Karst at Regional Scale (Berchtesgaden Alps, Germany)

Gabriele Kraller, Michael Warscher, Ulrich Strasser, Harald Kunstmann and Helmut Franz

Abstract Distributed hydrological modeling in karst dominated catchments is challenging as various unknown underground flow conditions and flow directions lead to unknown storage quantities. Missing parameterization in karst catchments at regional scale prevents reliable hydrological modeling of subsurface (unsaturated and saturated) water fluxes; and consequently, climate impact modeling in karst dominated catchments is until today insufficient. The deterministic hydrological model WaSiM-ETH by Schulla and Jasper was applied in the Alpine catchment of the river Berchtesgadener Ache to describe the water balance and to determine and quantify karst impact on hydrological processes at different time and space scales in the watershed. The study area is situated in the northern limestone Alps, characterized by a huge carbonate stratum, which is exposed to karstification processes since Alpine lift. It is assumed, that subsurface flow channels and heterogeneous storage effects lead to groundwater redistribution through mountain

G. Kraller (✉)
Department of Geoinformatics-Z_GIS, University of Salzburg, Hellbrunnerstrasse 34, 5020 Salzburg, Austria
e-mail: gabriele.kraller@sbg.ac.at

M. Warscher · H. Kunstmann
Institute for Meteorology and Climate Research (IMK-IFU), Karlsruhe Institute of Technology (KIT), Kreuzeckbahnstr. 19, 82467 Garmisch-Partenkirchen, Germany

U. Strasser
Institute for Geography, University of Innsbruck, Innrain 52f, 6020 Innsbruck, Austria

H. Kunstmann
Institute for Geography, University of Augsburg, Universitätsstr. 10, 86159 Augsburg, Germany

H. Franz
Berchtesgaden National Park Administration, Doktorberg 6, 83471 Berchtesgaden, Germany

J. Mudry et al. (eds.), *H2Karst Research in Limestone Hydrogeology*,
Environmental Earth Sciences, DOI: 10.1007/978-3-319-06139-9_8,

ranges and influence hydrological processes of neighboring valleys. In a first step, former karst research in the area is evaluated to draw the main subsurface flow directions within or in between sub-basins. Based on this, the water balance of the sub-basins is examined to obtain further information on the regional hydrology. This is done by analyzing model results of the hydrological model. A systematic mismatch between modeled and measured runoff (over and underestimation) was detected in three high Alpine karst dominated sub-basins, indicating hydrological subsurface processes at sub-basin scale. The comparison of monthly sums of modeled and measured water storage indicates subsurface water inflow, outflow or redistribution in sub-basins and enables quantification of those processes. Based on these outcomes, a method to predict future water storage in the Berchtesgaden karst is developed and groundwater modeling is adapted in WaSiM-ETH, which was developed to improve the hydrological model for karst-dominated catchments.

Keywords Alpine karst • Distributed hydrological modeling • Artificial neural network • Water balance

1 Introduction

Karst aquifers consist mainly of two components, the rock matrix and the conduits between (Atkinson 1977; Bakalowicz 2005; Kiraly 2003; Sauter et al. 2006; Teutsch and Sauter 1991; White 2002). White (2003) describes fractures as a third component of karst aquifers. This leads to a broad field of hydraulic parameters and unknown boundary conditions, which form the main challenge of describing the groundwater flow in the existing approaches (Kiraly 2003; White 2002; Sauter et al. 2006). Mountain massifs can be seen as lifted aquifers characterized by high altitudinal gradients and inclined stratification (Kraller et al. 2011). The head difference between recharge area and springs is the primary driving force for the movement of water through the aquifer (White 2003). One can assume that groundwater flow in Alpine catchments is increased. Being consistent with banked karstified limestone, the increased groundwater flow may even be additionally strengthened and redistribution processes may take place at the sub-basin scale and affect receiving water courses in neighboring Alpine valleys (Kraller et al. 2011). Many studies in karstified areas focus on small-scale effects of karst conduits by examining spring hydrographs, chemographs and tracer breakthrough curves—to define the size and the characteristics of one individual spring aquifer (Birk et al. 2004; Bonacci 1993; Einsiedl 2005; Geyer et al. 2008; Hauns et al. 2001; Kovács et al. 2005). Scanlon et al. (2003) applied a distributed groundwater model for discharge modeling of single wells. Those studies do not provide answers to questions about karst effects on the comprehensive water balance of an Alpine headwater catchment, i.e., if water redistributes from one sub-basin to another through flow cannels, and if groundwater flow in karstified mountains can be quantified (Kraller et al. 2011).

The distributed hydrological WaSiM-ETH (Schulla and Jasper 2012) is applied in the watershed of the Berchtesgadener Ache to describe the water balance and the impact of karst on the hydrology of the region. It is a modular hydrological model system that includes all relevant hydrological processes within the water cycle and provides outputs to all hydrological relevant processes within the area. As the groundwater module of the model does not account for karstic environments, the systematic model mismatch was analyzed to describe karst impact within and between neighboring Alpine sub-basins on the hydrology of the region. In a first step, previous karst research is analyzed to draw the main underground flow directions of the area. Then hydrological model results are analyzed and systematic mismatches are used to define redistributed water quantities within the karst aquifer at sub-basin scale. This knowledge is the basis for the development of a new method to correct distributed models for karst environments at regional scale.

2 Study Area

The test site for this study is situated in the Berchtesgaden Alps and covers an area of 432 km^2. It is characterized by extreme topography formed by single mountain ranges and ridges and the valleys in between, spanning 2,100 m in the vertical. Dominant rock formations are Triassic Dachstein limestone and Ramsau Dolomite, where the banked limestone comprises a layer thickness up to 1,000 m. The limestone was exposed to dissolution processes since the Cretaceous, leading to a massive karstified aquifer with a wide range of subsurface flow channels. There are hundreds of springs as groundwater recharge locations, feeding the seven rivers of the region. Generally the valleys are drained by rivers in the northern direction, and are associated with the Danube watershed. Three head sub-basins are situated in high Alpine karst terrain (Kraller et al. 2011).

3 Regional Hydrology

Former research in the basin is evaluated and summarized to describe groundwater flow to determine the main drainage direction, travel times, spring dynamics and possible subsurface redistribution in the single mountain ranges, and the whole basin. It aims to emphasize all outcomes indicating karst influence for the whole watershed or single sub-basins. Figure 1 summarizes the main outcomes. It was found that water flow is increased within sub-basins (sub-basin Königsseetal) but also that groundwater redistribution takes place (sub-basin Klausbachtal, Wimbachtal, Königsseetal). Furthermore, water outflow of the watershed is indicated (Fig. 1). The main subsurface drainage direction is north (Kraller et al. 2011).

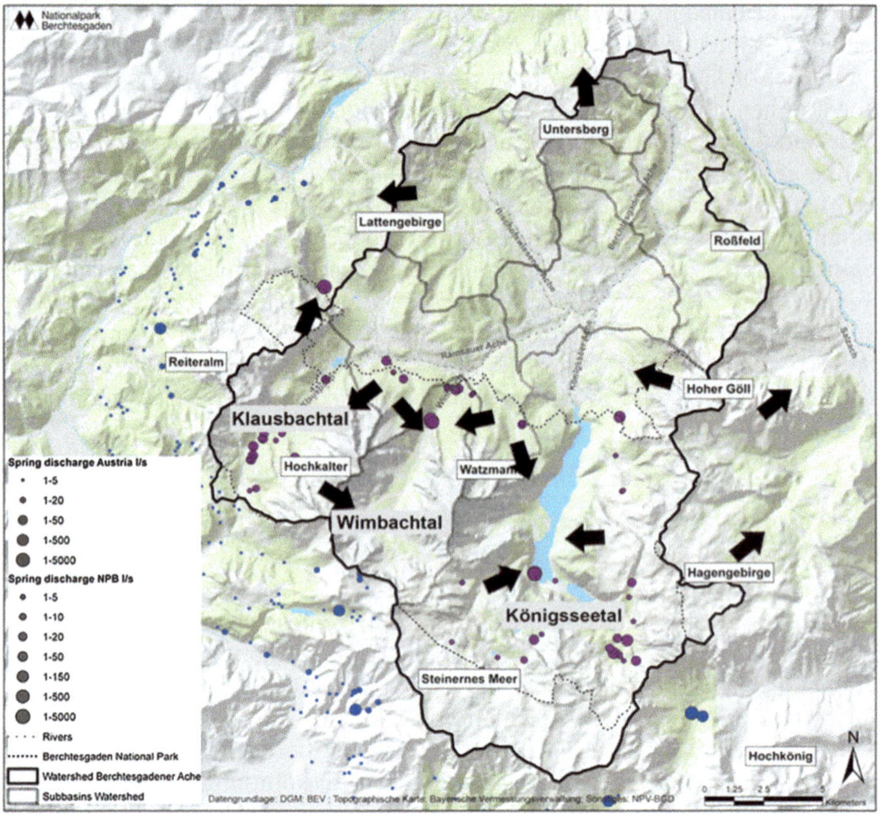

Fig. 1 Main subsurface water flow directions in the Berchtesgaden watershed (modified after Kraller et al. 2011)

4 Distributed Modeling

The water balance is modeled with the distributed model WaSiM-ETH (Schulla and Jasper 2012). It is a physically-based model that was applied in 50×50 m^2 horizontal resolution. WaSiM-ETH uses physically-based algorithms to calculate the majority of hydrological processes within the water balance, which are calculated one after another during model run for each grid cell and each time step (1 h). The model is applied from 2001 to 2010 whereby model calibration is done in 2001–2002 and model validation in 2002–2010. For the description of the soil water fluxes in the unsaturated zone, the solution of the Richards' equation (Richards 1931) is used. Groundwater flow is calculated with an integrated multilayer 2D-groundwater flow model, which couples dynamically with the unsaturated zone. The flux equation is derived from the continuity equation and Darcy-equation (Schulla and Jasper 2012). The model is not able to account for the karstic environment and assumes porous conditions. Next to numerous free model parameters, sensitive calibration parameters are the recession constant

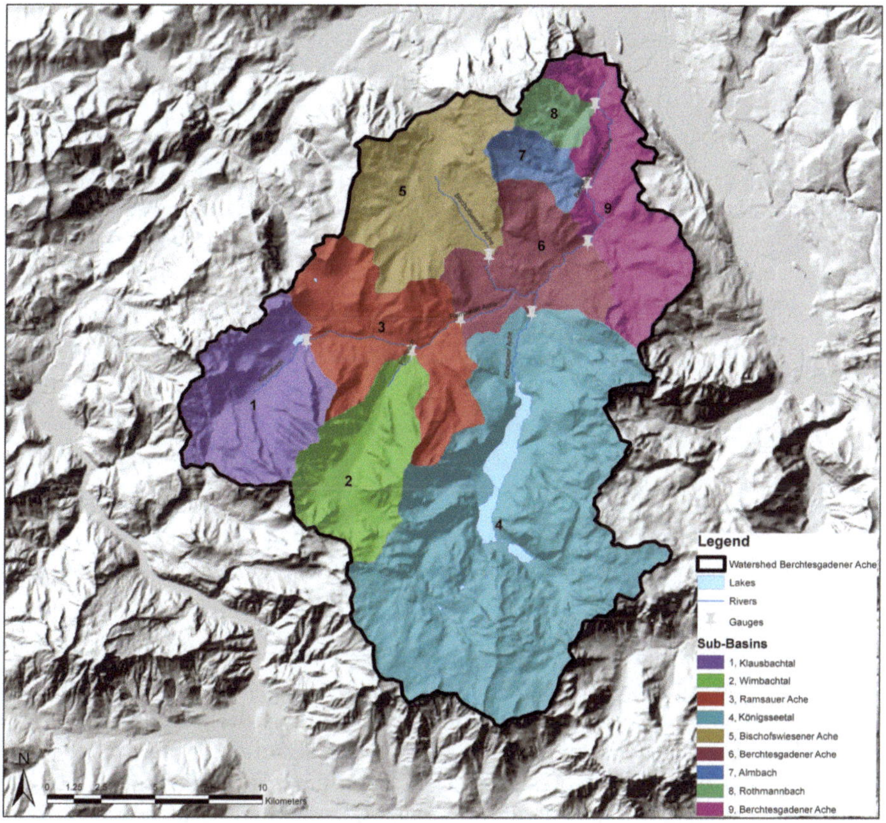

Fig. 2 Berchtesgaden watershed and sub-basins

k_{rec} for the saturated hydraulic conductivity and the interflow drainage density d_r (Schulla and Jasper 2012). Model input data were derived from several sources: soil data and land use data is derived from the National park database. Data for the hydraulic conductivity is derived from the hydrogeological map of Bavaria and is assumed to decrease with depth according to a recession constant dependent on the soil texture. Meteorological data was used from automatic and non-automatic climatic stations within the area. Based on available river gauges and the digital elevation model, the area can be separated into nine sub-basins (Fig. 2).

5 Model Results

First, the authors found that annual sums of measured runoff significantly differ between the high Alpine and mostly karst influenced head sub-basins Klausbachtal, Wimbachtal, Königsseetal (Fig. 1). When comparing measured runoff in the three high Alpine sub-basins, Wimbachtal and Königsseetal exceed

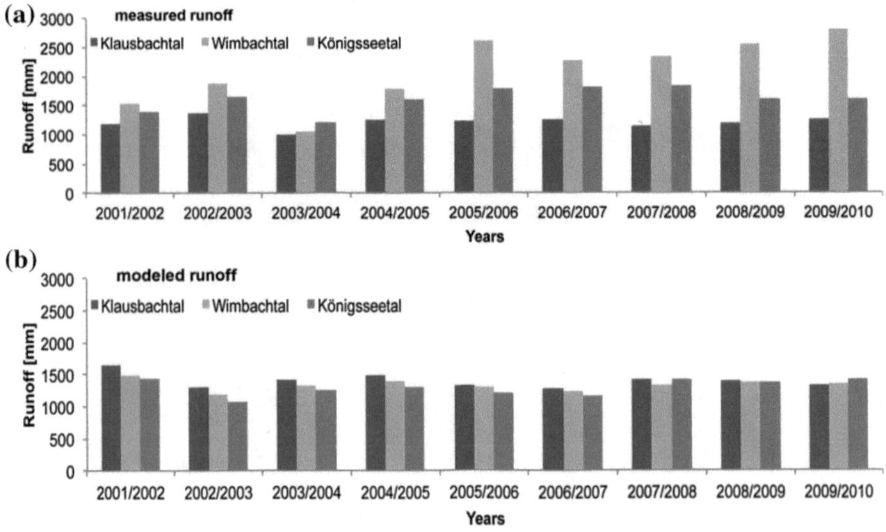

Fig. 3 Annual sums of modeled (**b**) and measured discharge (**a**) sub-basins Klausbachtal, Wimbachtal and Königsseetal

sub-basin Klausbachtal distinctly (Fig. 3). After model calibration, the modeled runoff results in equally calculated annual sums, dependent on precipitation input. Consequently, a systematic over and underestimation of discharge was found in these sub-basins when comparing modeled to measured runoff (Fig. 3). Runoff in sub-basin Klausbachtal is overestimated and runoff in sub-basin Wimbachtal and Königsseetal is underestimated by the model. The authors assume that different annual sums of measured runoff in the three sub-basins could be a result of subsurface water inflow, outflow or redistribution through karst conduits in this high Alpine karst aquifer. This is not taken into account by the hydrological model, leading to under and overestimation of measured runoff during model runs.

In a next step, monthly sums of water storage S(t) (Eq. 1) were analyzed for sub-basins Klausbachtal, Wimbachtal and Königsseetal in model runs and reality to gain more information about the annual dynamics of measured runoff and the underlying reasons of model mismatch.

$$S(t) = NS_{eff}(t) - Q(t) \tag{1}$$

By subtracting the modeled and measured runoff (Q) from the modeled effective precipitation (NS_{eff} = precipitation − evapotranspiration), storage reduction or buildup is expressed and is assumed to be positive in winter and summer (snow storage and soil storage) and negative in spring and autumn (snow melt and soil storage decrease), leading to a systematic pattern throughout a year leading to an even water balance equation. By comparing modeled to measured water storage, deviations of the assumed pattern for the water storage may give insights for groundwater inflow, outflow or redistribution at sub-basin scale due to karst conduits within one hydrological year. Figure 4 shows the observed and modeled

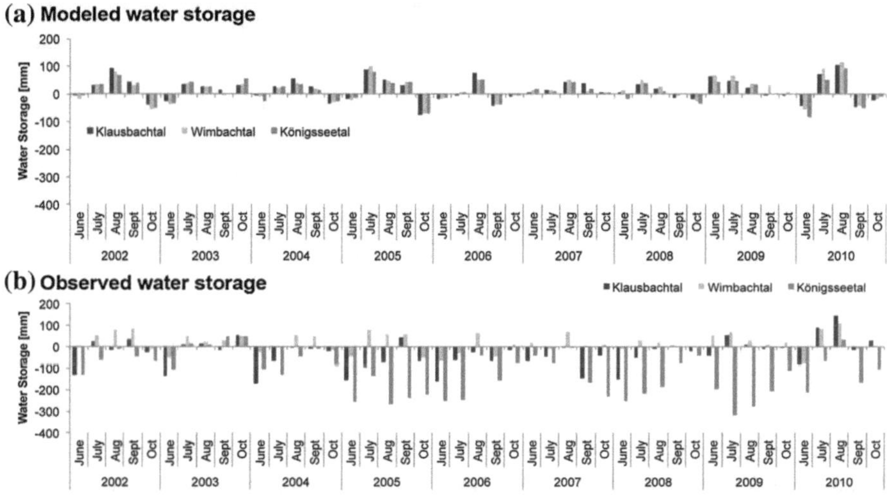

Fig. 4 **a** Comparison of the water storage (Smod) derived from results of hydrological model runoff. Monthly sums June–October 2002–2011. Sub-basins Klausbachtal, Wimbachtal, Königsseetal. **b** Comparison of the water storage (Sobsreal) derived from measured runoff. Monthly sums June–October 2002–2011. Sub-basins Klausbachtal, Wimbachtal, Königsseetal (modified after Kraller et al. 2012)

water storage in monthly sums for the year 2002–2010 for the analyzed sub-basins. The observed water storage is derived from the measured runoff and the modeled water storage is derived from modeled runoff.

The analysis of the observed water storage (Fig. 4b) shows that during spring and summer pronounced negative peaks occur, indicating groundwater inflow and explaining the amount of measured annual runoff. In sub-basin Klausbachtal there is less soil storage and snow melt, indicating groundwater outflow and explaining the annual lack of water. The water storage calculated by the model (Fig. 4a) differs from observed water storage in spring and summer months. It is assumed, that due to groundwater inflow or outflow real catchment sizes are larger or smaller than are presumed for distributed modeling leading to systematic mismatch of modeled and measured runoff at the end. These results coincide with the assumed regional hydrology (Kraller et al. 2011).

6 Adaptation of the Distributed Model to Karst Environments

6.1 Artificial Neural Network and Distributed Model Correction

The preceding steps show that the hydrology of the three high Alpine neighboring sub-basins Klausbachtal, Wimbachtal and Königsseetal is unique for each and that the hydrological model is not able to reproduce such conditions. These limitations

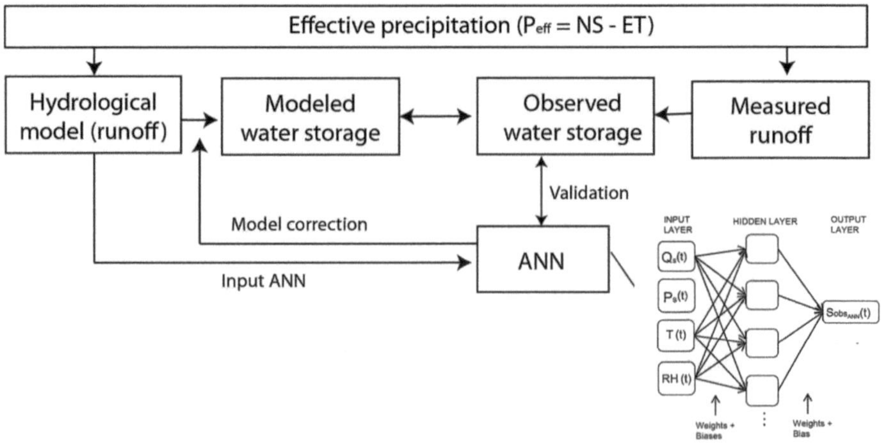

Fig. 5 Overview of the presented method. Effective precipitation results in modeled and measured runoff. Consistent model mismatch was detected due to water storage deviations. Modeled water storage (derived from hydrological model runoff) systematically over/underestimates observed water storage (derived from measured runoff). The observed water storage is calculated by the ANN and implemented in the groundwater module to account for the observed storage processes. The ANN is applied to the subbasin Königsseetal (Kraller 2012). The artificial neural network consists of an input layer, hidden layer and output layer. The inputs QS(t) (snow melt and/or rain), snow precipitation PS(t), temperature T(t), relative humidity RH(t) give the observed water storage SobsANN(t) (modified after Kraller et al. 2012)

cannot be explained by precipitation interpolation methods or precipitation correction (Kraller et al. 2012). The measured runoff in those sub-basins is attributed to very heterogeneous subsurface water storage and flux conditions in karst terrain. It is concluded, that underestimated water quantities in the model are a consequence of subsurface water storage or influx processes within each of the sub-basins. Only boundary fluxes from outside the watershed boundary into the sub-basin are assumed. Therfore, the hydrological model needs to be adapted. The aim was to find an adequate method to quantify storage conditions in the validation period of the hydrological model WaSiM-ETH, to reproduce them and to find a possibility to account for the storage also within the hydrological model (Fig. 5). The target to be reproduced is the water inflow and outflow into the karst system, which can be expressed as the observed water storage. This water storage in the karst aquifer is temporally very diverse and underlies many influencing processes such as precipitation quantity and intensity, evapotranspiration and soil water storage. These processes and their interaction are highly complex, and it is not possible to physically describe those in detail within the study area.

To learn and determine the observed water storage in highly karstified watersheds, an Artificial Neural Network (ANN) was developed, trained and applied. ANNs are nonlinear input-output models which are able to reproduce the desired output based on a given input parameter set. An ANN is able to determine complex input-output relationships where physically-based methods are limited. The main advantage of this method is that the network itself is trained by

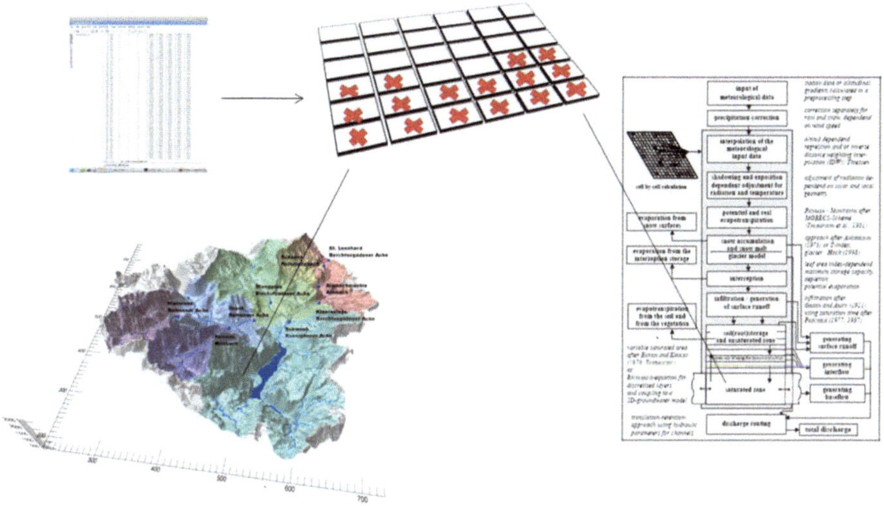

Fig. 6 Technical implementation of the influx in the distributed model (modfied after Kraller et al. 2012)

parameter adjusting and after validation available itself as tool for calculating subsurface conditions. Due to the algorithm, negative and positive water storage can be considered. The ANN aims to quantify the observed water storage for later correcting the hydrological model. In this study, a two-layer feedforward backpropagation network was applied. The architecture of the ANN is shown in Fig. 5. Input variables for the ANN are hydrological model outputs. The output variable is the observed water storage, or in other words, unknown subsurface boundary water fluxes. The water storage calculated by the ANN is then converted into a temporal dynamic influx into or outflow of the groundwater module of the hydrological model. It is used for hydrological model correction in the saturated zone to adapt the model to karstic environments (Fig. 6). This new method is developed, applied and tested in the high Alpine sub-basin Königsseetal. The implemented influx is then considered in the consecutive sub-basins (Berchtesgadener Ache and St. Leonhard). Monthly sums of modeled discharge and also runoff dynamics at three river gauges (Königsseetal, Berchtesgadener Ache and St. Leonhard) are analyzed to validate the method (Kraller et al. 2012).

7 Results

The development of the new method shows that it is capable to reproduce missing water fluxes in karst terrain. The missing water fluxes are then converted in a boundary flux that is applied in the distributed model (Fig. 6). The performance of the model correction is shown in Fig. 7 for the sub-basins Königsseetal and the downstream sub-basins Berchtesgadener Ache and St. Leonhard. Monthly

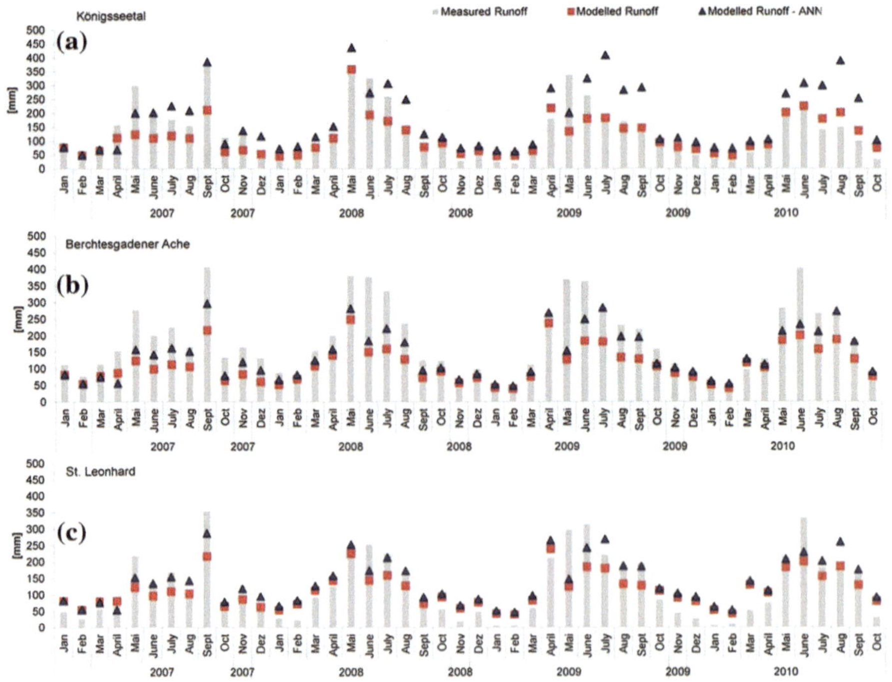

Fig. 7 Monthly sums of measured runoff and hydrological model runoff with and without implemented boundary flux from June to October for years 2007–2010 for sub-basins Königsseetal (**a**), Berchtesgadener Ache (**b**) and St. Leonhard (**c**) (modified after Kraller et al. 2012)

sums of modeled water storage with and without boundary flux are compared to monthly sums of measured runoff. The analyzed time period is the test period of the ANN (2007–2010). The method leads to an increasing performance in sub-basin Königsseetal in the years 2007 and 2008 where there is a great discrepancy between measured and modeled runoff in summer months. In the years 2009 and 2010 the model correction is overestimating measured runoff, because the deviation of modeled and measured runoff before correction was already not as big as the years before. In sub-basins Berchtesgadener Ache and St. Leonhard, the method is increasing hydrological model performance. Results of the application of the consequent flux in the hydrological model are discussed in detail in Kraller et al. (2012).

8 Conclusion

The hydrological model was applied in the catchment of the river Berchtesgadener Ache to analyze the water balance within the area and to describe possible karst impact on regional hydrology at sub-basin scale. As the hydrological model does not account for karstic environment, systematic model mismatch indicates possible subsurface inflow, outflow and redistribution between sub-basins within the high Alpine environment. Analyses of monthly sums of observed and modeled water storage was conducted to gain more knowledge about observed and modeled storage dynamics. Based on this a new method to adapt the hydrological model to karst environments was developed. An Artificial Neural Network (ANN) is capable to reproduce and quantify missing water fluxes. This enables the implementation of a continuous inflow/outflow within the groundwater module (boundary flux) in the distributed model to account for missing subsurface water fluxes. The performance of the hydrological model is then increased. Furthermore, it is possible to apply the ANN in scenario calculations to ensure more realistic distributed modeling in karst terrains in climate impact runs.

References

Atkinson T (1977) Diffuse flow and conduit flow in limestone terrain in the Mendip Hills, Somerset (Great Britain). J Hydrol 35:93–110

Bakalowicz M (2005) Karst groundwater: a challenge for new resources. Hydrogeol J 13:148–160

Birk S, Liedl R, Sauter M (2004) Identification of localised recharge and conduit flow by combined analysis of hydraulic and physico–chemical spring responses (Urenbrunnen, SW-Germany). J Hydrol 286:179–193

Bonacci O (1993) Karst springs hydrographs as indicators of karst aquifers. Hydrol Sci J 38:1, 2

Einsiedl F (2005) Flow system dynamics and water storage of a fissured-porous karst aquifer characterized by artificial and environmental tracers. J Hydrol 312:312–321

Geyer T, Birk S, Liedl R, Sauter M (2008) Quantification of temporal distribution of recharge in karst systems from spring hydrographs. J Hydrol 348:452–463

Hauns M, Jeannin P-Y, Atteia O (2001) Dispersion, retardation and scale effect in tracer breakthrough curves in karst conduits. J Hydrol 241(3):177–193

Kiraly L (2003) Karstification and groundwater flow. Speoleogenesis Evol Karst Aquifers 155–190

Kovács A, Perrochet P, Király L, Jeannin P-Y (2005) A quantitative method for the characterisation of karst aquifers based on spring hydrograph analysis. J Hydrol 303:152–164

Kraller G (2012) Distributed modeling of the water balance in a high Alpine karstified watershed (Berchtesgaden Alps, Germany). Development of an artificial neural network extension to account for dynamic karst aquifer water storage at subbasin scale. Dissertation, Karl-Franzens Universität Graz 2012, 174 S

Kraller G, Strasser U, Franz H (2011) Effect of Alpine karst on the hydrology of the basin "Berchtesgadener Ache": a comprehensive summary of karst research in the Berchtesgaden Alps. Eco. mont 3(1):19–28

Kraller G, Warscher M, Kunstmann H, Vogl S, Marke T, Strasser U (2012) Water balance estimation in high Alpine terrain by combining distributed modeling and a neural network approach (Berchtesgaden Alps, Germany). Hydrol Earth Syst Sci 16:1969–1990

Richards L (1931) Capillary conduction of liquids through porous mediums. Physics 1:318–333

Sauter M, Kovacs A, Geyer T, Teutsch G (2006) Modellierung der Hydraulik von Karstgrundwasserleitern—Eine Übersicht. Grundwasser 3:143–156

Scanlon B, Mace R, Barrett M, Smith B (2003) Can we simulate regional groundwater flow in a karst system using equivalent porous media models? Case study, Barton Springs Edwards aquifer, USA. J Hydrol 276:137–158

Schulla J, Jasper K (2012) Model description WASIM-ETH (water balance simulation model ETH). ETG-Zürich, Zürich
Teutsch G, Sauter M (1991) Groundwater modeling in karst terranes: scale effects, data acquisition and field validation. In: Proceedings of 3rd conference on hydrogeology, ecology, monitoring and management of ground water in karst terranes, S. Nashville
White WB (2002) Karst hydrology: recent development and open questions. Eng Geol 65:85–105
White WB (2003) Conceptual models for karstic aquifers. Speleogenesis Evol Karst Aquifers 1(1):3

Diagnostic Plots Applied to Well-Tests in Karst Systems

Jean-Christophe Maréchal, Bernard Ladouche, Benoît Dewandel, Perrine Fleury and Nathalie Dörfliger

Abstract Pumping tests conducted on wells intersecting karst heterogeneities, such as the conduit network, are difficult to interpret. Nevertheless, this case can be solved by assimilating the horizontal karst conduit to a finite-conductivity vertical fracture. In this case, several flow patterns corresponding to the respective contributions of karst subsystems (fractured matrix, small conduits and main karst drainage network) can be identified on the diagnostic plot of drawdown derivative. This is illustrated on two examples from Mediterranean karst systems. A pumping test on a well crossing the main karst drainage network of the Cent-Fonts karst system shows: (1) a preliminary contribution of the karst conduit storage capacity followed by (2) linear flows into the fractured matrix. A pumping test on a well intersecting a small karst conduit of the Bas-Agly karst system shows the existence of (1) bi-linear flows within both the karst conduit an the fractured matrix at early times, followed by (2) radial flows within the fractured matrix and (3) finally the contribution of a major karst cavity. The use of diagnostic plots allows the identification of the various flow regimes during pumping tests, corresponding to the response of the individual karst aquifer subsystems. This is helpful in order to understand the structure of the karst aquifer and flow exchanges between subsystems.

Keywords Mediterranean karst systems • Karst drainage network • Karst conduit storage capacity • Pumping tests • France

J.-C. Maréchal (✉) · B. Ladouche · B. Dewandel · P. Fleury · N. Dörfliger
BRGM, D3E/NRE, 1039 rue de Pinville, 34000 Montpellier, France
e-mail: jc.marechal@brgm.fr

J. Mudry et al. (eds.), *H2Karst Research in Limestone Hydrogeology*,
Environmental Earth Sciences, DOI: 10.1007/978-3-319-06139-9_9,

1 Introduction

Due to the complexity and duality of flows, well-test interpretation in karst systems constitutes a challenging task for hydrogeologists (Kresic 2007). This is especially true when the pumping well intersects karst heterogeneities such as the conduit network for example. Nevertheless, well-tests can bring valuable information on the presence and size of karst heterogeneities.

Diagnostic plots of drawdown derivative have been widely used in oil industry for fractured reservoirs characterization. These techniques are now used by hydrogeologists (Maréchal et al. 2004; Renard et al. 2009) in order to establish conceptual models of fractured aquifers under pumping. This method can be applied to karst hydrogeology as well. In this chapter, the classical response of a well test into a large fracture (or karst conduit) is described on a log–log drawdown derivative curve. It allows identifying successive flow regimes corresponding to the contribution of various karst aquifer subsystems (fractured matrix within the vadose zone, karst conduit, main karst drainage network) to the pumped well. The interpretation of such a diagnostic plot is illustrated on two examples from Mediterranean karst systems.

2 Method

In heterogeneous karst systems, the log–log diagnostic plot of drawdown and its derivative in the pumping well or in observation wells can help in identifying departures in flow-geometry from the classical homogeneous radial case. The common logarithm of drawdown s change is plotted versus the logarithm of elapsed time t, together with log ($\mathrm{d}s/\mathrm{d} \ln t$), the logarithm of the derivative of drawdown change with respect to the natural logarithm of time. Several authors (Ehlig-Economides 1988) noticed that the drawdown derivative displays late-time straight lines, with slopes related to the domain dimensionality and boundary conditions. Classically, the log–log diagnostic plot can be divided into several portions: (a) early data used for identifying the wellbore or karst conduit storage; (b) intermediate data for identifying the type of aquifer model that should be used (e.g. double porosity, unconfined, drainage); and (c) late data for identifying the possible boundaries.

Siting a pumping well on the karst conduit network (Fig. 1) of a karst system constitutes an interesting target regarding further possible karst aquifer exploitation. This position of the pumping well allows draining of the highly permeable karst conduit which will then interact with the surrounding matrix. The karst conduit constitutes then a useful extension of the pumping well. Two separate cases are distinguished: (a) the case of a pumping well intersecting the main karst drainage network connected to the outlet (PWa in Fig. 1) and (b) the case of a pumping well intersecting a small karst conduit (PWb in Fig. 1).

These cases can be compared to wells stimulated by hydraulic fracturing techniques as described in the petroleum literature. Consider a vertically fractured well producing at a constant flow rate in an infinite, isotropic, homogeneous horizontal

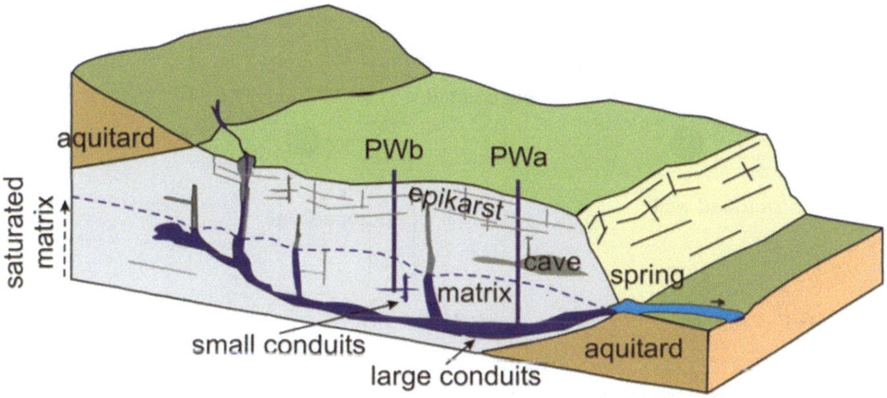

Fig. 1 Karst aquifer with a pumping-well (*PW*) located into the karst conduit network (modified from Goldscheider and Drew 2007): PWa pumping well crossing the main karst drainage network; PWb pumping well crossing a small karst conduit

aquifer. Figure 2 shows the general behaviour of a well with a horizontal conduit that can be assimilated to a finite-conductivity vertical fracture. The flow-pattern geometry under such conditions will change with time and includes several flow periods (Cinco-Ley and Samaniego-V 1981). During early pumping times (Fig. 2a), conduit linear flow (conduit or wellbore storage effect) is characterized by a straight line with a slope $\nu = 1$ on the diagnostic plot. After a transition flow period, the system may or may not exhibit a bi-linear flow period (Fig. 2b), indicated by a one-fourth-slope straight line that lasts as long as the conduit boundaries do not affect the flow pattern. As time increases, the flow in the matrix becomes one-dimensional (horizontal, parallel and perpendicular to the conduit) as illustrated on Fig. 2c. On a diagnostic plot, the slope of the derivative curve is $\nu = 0.5$. As pumping continues, the flow pattern can change from parallel flow to pseudo-radial flow (Fig. 2d). During this period, the pumped water originates from farther away and the local effect of the finite-length conduit tends to disappear. The drawdown curve tends to the Jacob straight line with a slope $\nu = 0$ of the derivative at late times; however, this would arise only if the duration of the pumping test is very long. This four-period flow-pattern geometry makes it impossible to interpret such drawdown curves with classical techniques. The corresponding typical drawdown and derivative curves (diagnostic plot) are illustrated on Fig. 2e.

3 Study Areas

The Cent-Fonts karst (Fig. 3) is a mixed-flow karst system located north of Montpellier (Hérault region, Southern France) in a thick limestone and dolomite series (Middle and Upper Jurassic). It has been characterized by long-term

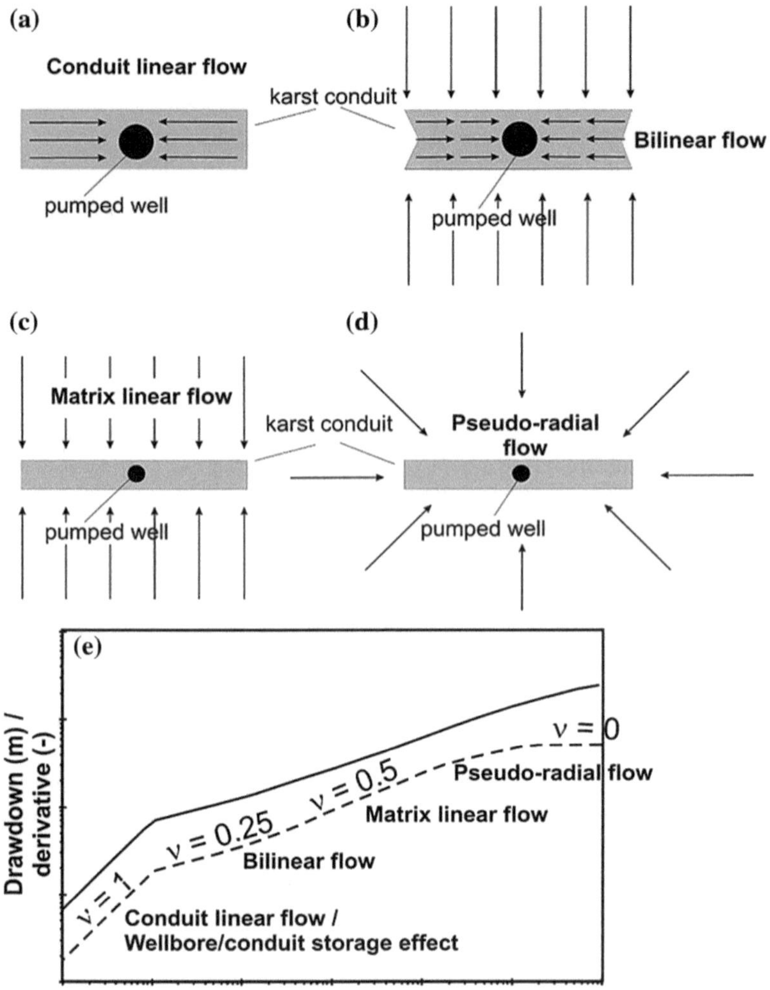

Fig. 2 A well intersecting a karst conduit acting as a vertical fracture of finite length and finite hydraulic conductivity: **a** linear flow in the conduit or wellbore/conduit storage effect during early pumping times; **b** bi-linear flow; **c** matrix linear flow; **d** matrix pseudo-radial flow during late pumping times, modified after (Cinco-Ley and Samaniego-V 1981); **e** diagnostic plot of the same case

monitoring (1997–2007) and numerous studies (Ladouche et al. 1999, 2002, 2005, 2006; Aquilina et al. 1999, 2005, 2006). The Cent-Fonts spring, located on the right bank of the Hérault River, is the only outlet of the karst system (Fig. 3a). Its discharge ranges from Q_S = 220 L/s (during severe low water stage periods) to more than 12,000 L/s during peak flow in winter or spring, the average spring discharge having been about 1 m^3/s during the 1997–2005 period (Ladouche et al. 2005). The Cent-Fonts spring is the outlet of a water-saturated karst conduit

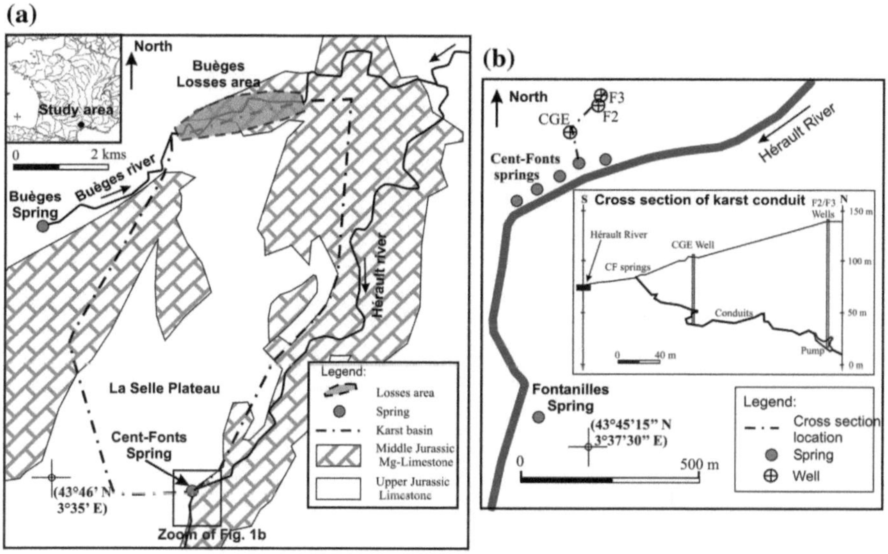

Fig. 3 **a** Geological and location map of the cent-fonts karst system. **b** Location of pumping well (F3); cross section showing the wells intersecting the conduit network

network that has been partially explored and mapped by divers in the vicinity of the spring, particularly for the purpose of the pumping well sitting (Bardot 2001 in Ladouche et al. 2005). The network is developed below the Hérault River to a depth of at least 95 m (Fig. 3b). In the mapped area, the cross-sectional area of the karst drain ranges from 4 to 16 m^2 and its largest part (a 400 m^3 cavity) is located at the end of the explored karst conduit (Fig. 3b). Three wells intercept the karst network near the spring (Fig. 3b): CGE is about 60 m deep, F3 is located about 100 m upstream from CGE and reaches the largest part of the conduit at a depth of 128 m, and F2 is located 3 m from F3 in the same conduit.

The Bas-Agly karst system is located on the Mediterranean coast in Southern France (Pyrénées Orientales), near the Salses-Leucate lagoon (Dorfliger et al. 2009). Due to the complex tectonics of the Corbières, strongly influenced by Pyrenean orogenesis, the Jurassic-Cretaceous limestone aquifer is not a continuous aquifer. The depth of karstification, several hundreds of meters below sea level, is linked to the Messinian Salinity Crisis. This system is partially cut off from the Mediterranean Sea by the Plio-Quaternary sedimentary cover. To the north, the karst is in contact with the brackish Salses-Leucate lagoon (Fig. 4). The two main springs of this karst system are the Font Estramar spring and the Font Dame spring, which emerge near the brackish lagoon, a few kilometers from each other. The Font Dame spring is characterized by several outlets in a wetland with a low water stage discharge of 300 L/s and a high water stage discharge of 6,000 L/s. The Font Estramar spring is located at the foot of a Cretaceous limestone cliff. Its discharge is between 800 L/s and more than 20,000 L/s. The total mean annual flow rate for the two springs is 2,500 L/s (1,700 L/s at Font Estramar and 800 L/s at Font Dame).

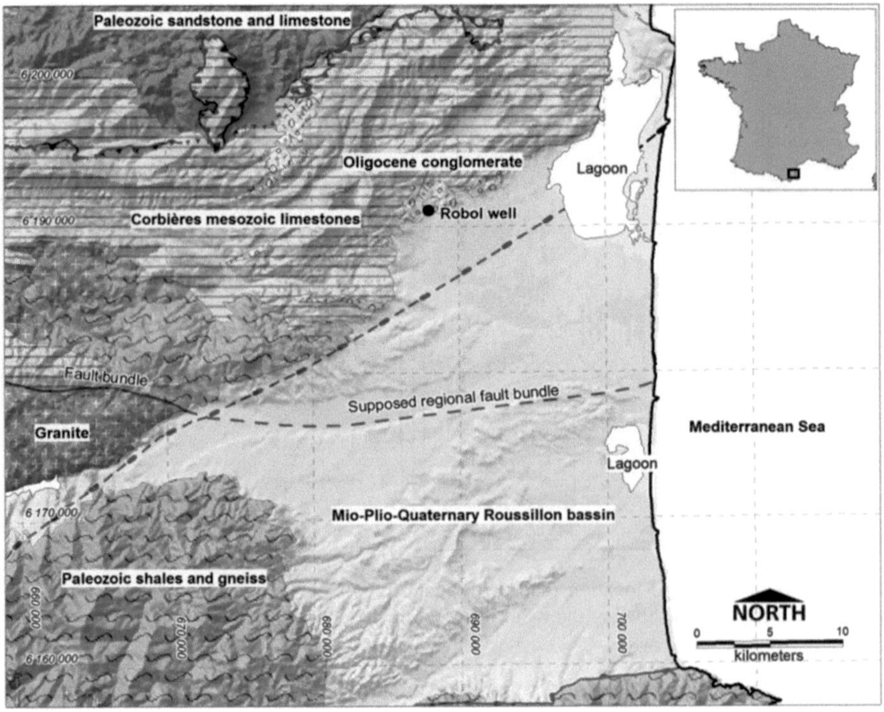

Fig. 4 Geological and location map of the Bas-Agly karst system and Robol well

4 Results

4.1 Cent-Fonts Karst System

A long-duration pumping test (1 month) was conducted on the main conduit of the Cent-Fonts karst system (Marechal et al. 2008, Ladouche et al. 2007). The well F3 has been sited by diving exploration in order to intersect the main karst conduit about 100 m upstream from the spring of the system (Fig. 1b). The well was pumped at a constant discharge rate of 400 L/s. During the first 24 h ($\approx$1,500 min), the storage effect in the karst conduit prevailed (slope $\nu = 1$, Fig. 5a). After that, the drawdown tended to linear flow with a slope decreasing to $\nu = 0.7$. No bi-linear flow period was observed. In fractured systems, Cinco-Ley and Samaniego-V (1981) consider that this behaviour is not present when the fracture has a high storage capacity and a high conductivity. This is the most probable reason for such behaviour, given the large diameter (3.5 m) of the karst conduit in the vicinity of the pumping well.

Analytical solutions for a pumping well intersecting a vertical fracture of infinite conductivity can be transposed to a well intersecting a high-conductivity karst conduit. Such solutions were developed by petroleum engineers for application to

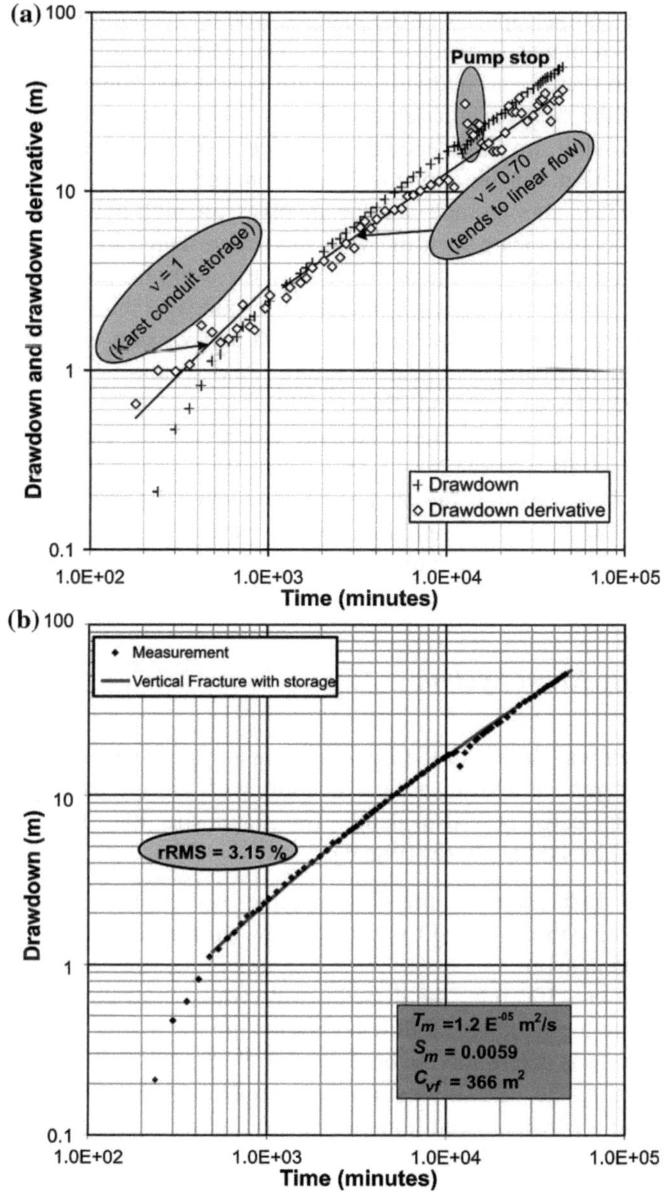

Fig. 5 **a** Diagnostic plot of the cent-fonts pumping test; **b** temporal evolution of measured drawdown and fitting with the single-vertical-fracture method of (Ramey and Gringarten 1976), modified after (Maréchal et al. 2008)

artificially fractured wells (Kruseman and Ridder 1990). They deal with a fully penetrating, vertically fractured well pumped at a constant rate, in an ideal, homogeneous, and horizontal formation (corresponding to the matrix in the karst system) of constant thickness, porosity (S_m) and permeability (T_m).

In case the storage capacity of the pumped well and conduit is low, the fracture is idealized as a planar surface of zero thickness and infinite hydraulic conductivity, such that there is no hydraulic-head drop as water flows internally through the fracture to the well (Gringarten and Ramey 1974a). In addition, if the well or the pumped conduit are large enough to induce a significant well-storage effect, a modified solution can be applied (Gringarten and Ramey 1974b; Ramey and Gringarten 1976). This storage volume involves all high-hydraulic conductivity volumes that communicate with the well. It is represented by C_{vf}, defined as the ratio between the change in volume of water in the well plus karst conduit, and the corresponding drawdown. This parameter corresponds to the free-surface area of the dewatering conduit network (vertical shafts and variably saturated conduits). Although cited in hydrogeological literature, field examples of these analytical solutions are rare (Kruseman and Ridder 1990).

The fit of the vertical fracture with the storage model on measured data is quite good after 10 h (rRMS = 3.15 %, Fig. 5b). Using the hypothesis of a 5,000-m-long karst conduit, the obtained hydrodynamic parameters are given on Fig. 5b. During this test, no pseudo-radial flow period was observed. This is most probably due to the great length of the karst network.

4.2 Bas-Agly Karst System

The Robol well was drilled in the Bas-Agly karst system near Perpignan (Fig. 4). With a total depth of 500 m, a karst conduit with an approximate diameter of 10 cm was intersected at 420 m depth. The diagnostic plot of the long-duration pumping test shows three phases (Fig. 6):

- During the first 20 min, the flow regime was characterized by a slope $\nu = 0.2 - 0.3$ corresponding to bi-linear flow (linear flow in the conduit and in the fractured matrix);
- Between 20 and 100 min, the drawdown derivative was constant ($\nu = 0$) corresponding to a pseudo-radial flow regime within the matrix. The transmissivity of the matrix was determined to be 2.5×10^{-4} m^2/s;
- After 100 min, the derivative strongly decreased because of stabilized drawdown due to a constant head boundary condition most probably corresponding to another large karst volume.

No conduit linear-flow regime was observed at the beginning of the test because the radius of the intersected karst conduit is small and its storage effect is therefore not measurable. The early transition to a pseudo-radial flow regime is due to the limited length of the karst conduit compared, for example, to the Cent-Fonts karst conduit. Therefore, no linear flow within the fractured matrix is observed.

These changes in flow regimes induce fluctuations in the electrical conductivity (EC) of the pumped water (Fig. 6). Let us assume that the pumped water results from the mixing of two end-members: less mineralized (low EC) conduit water

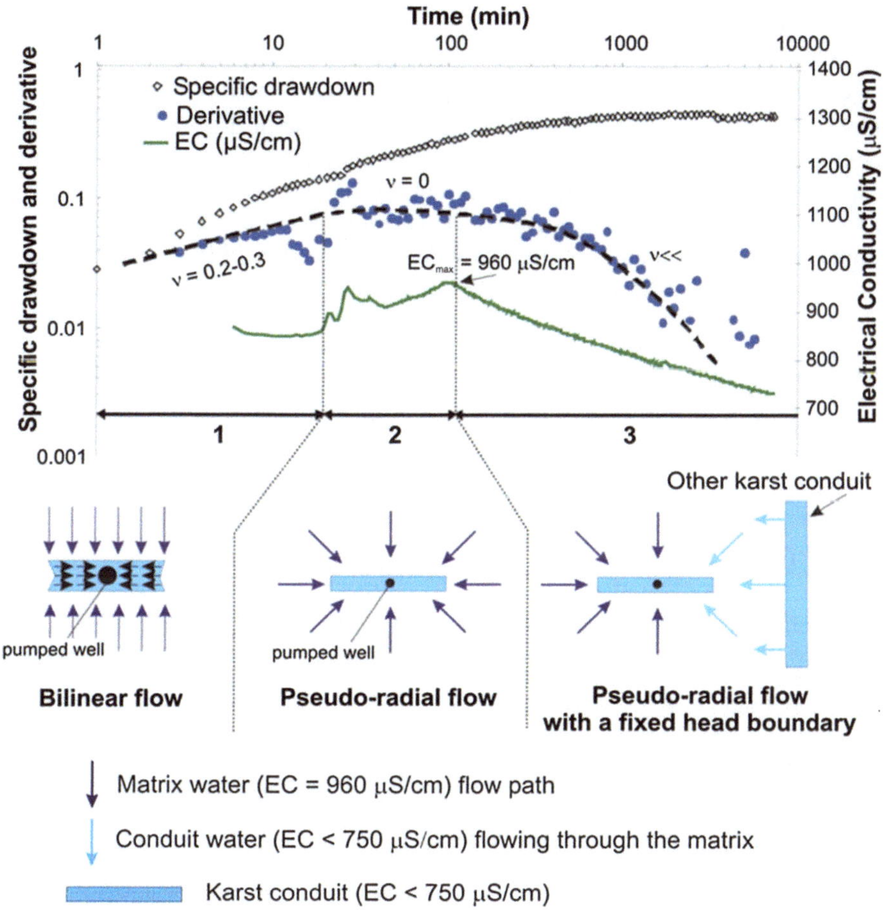

Fig. 6 Pumping test in the Robol well, intersecting a deep small karst conduit

and more mineralized (high EC) matrix water. During the first phase, the EC is average because the water is a mix of conduit and matrix water. During the second phase, the EC increases as the matrix contribution increases: the maximum value reaches ECmax = 960 µs/cm. Once the effect of another karst conduit is observed on the drawdown derivative (Phase 3), the water EC decreases toward that of the karst conduit water EC (≤750 µs/cm). The rapid EC decrease suggests the existence of a direct relationship between two karst volumes. This could be either a fracture network between two separate karst conduits, or a constriction separating a single karst conduit in two parts. Interestingly, the sharp EC increase of the water after 20–25 min induced by increased pumping rates was due to matrix-water inflow; the small pumped karst conduit could not provide the necessary water, which consequently flowed in from the matrix.

5 Conclusion

Classical methods for pumping tests interpretation cannot be applied to wells intersecting karst heterogeneities. Nevertheless, the use of diagnostic plots allows identification of various flow regimes during pumping tests, corresponding to the response of the individual karst aquifer subsystems (fractured matrix, small conduits, and main karst drainage network). The succession of various flow regimes brings information about the complexity of flow processes within the karst system. This is helpful in order to understand the structure of the karst aquifer and flow exchanges between subsystems.

Acknowledgments This research project was funded by BRGM as part of the Corbières Phase 3 and EAUR15-R2EAU projects. The General Councils of Aude and of Pyrénées Orientales, the Regional Council of Languedoc-Roussillon and Rhône, and the Mediterranean and Corsica Water Agency (AERM&C) also provided financial support.

References

Aquilina L, Ladouche B, Bakalowicz M, Schoen R, Pételet E (1999) Caractérisation du fonctionnement des systèmes karstiques nord-montpelliérains—Volume de synthèse générale, Report R40746, BRGM, Montpellier, France, 50 p

Aquilina L, Ladouche B, Dörfliger N (2005) Recharge processes in karstic systems investigated through the correlation of chemical and isotopic composition of rain and spring-waters. Appl Geochem 20:2189–2206

Aquilina L, Ladouche B, Dörfliger N (2006) Water storage and transfer in the epikarst of karstic systems during high flow periods. J Hydrol 327:472–485

Bardot (2001) Cartographie de la galerie noyée des Cent-Fonts. BARDOT & Co Report n 2001-09

Cinco-Ley H, Samaniego-V F (1981) Transient pressure analysis for fractured wells. J Pet Technol 33(9):1749–1766

Dörfliger N, Fleury P, Ladouche B (2009) Inverse modeling approach to allogenic karst system characterization. Ground Water 47(3):414–426

Ehlig-Economides C (1988) Use of the pressure derivative for diagnosing pressure-transient behavior. J Pet Technol:1280–1282

Goldscheider N, Drew DP (2007) Methods in karst hydrogeology. Taylor & Francis Group, London

Gringarten AC, Ramey HJ (1974a) Unsteady-state pressure distribution created by a well with a single horizontal fracture, partially penetrating or restricted entry. Trans Am Inst Min Eng 257:413–426

Gringarten AC, Ramey HJ (1974b) Unsteady-state pressure distributions created by a well with a single infinite-conductivity vertical fracture. Soc Pet Eng J:347–360

Kresic N (2007) Hydraulic methods. In: Goldscheider N, Drew D (eds.) Methods in karst hydrogeology. Taylor & Francis Group, London

Kruseman GP, Ridder NA (1990) Analysis and evaluation of pumping test data. ILRI Publication, Wageningen, 377 pp

Ladouche B, Aquilina L, Petelet E, Bakalowicz M, Schoen R (1999) Caractérisation du fonctionnement des systèmes karstiques nord-montpelliérains—Interprétation des données hydrochimiques. Report R 40940, BRGM, Montpellier, France, 170 p

Ladouche B, Dörfliger N, Pouget R, Petit V, Thiery D, Golaz C (2002) Caractérisation du fonctionnement des systèmes karstiques nord-montpelliérains—Rapport du programme 1999–2001-Buèges. Report BRGM/RP-51584-FR, BRGM, Montpellier, France, 200 p

Ladouche B, Maréchal JC, Dörfliger N, Lachassagne P, Lanini S, Le Strat P (2005) Pompages d'essai sur le système karstique des Cent-Fonts (Cne de Causse de la Selle, Hérault)—Présentation et interprétation des données recueillies. Report BRGM/RP 54426-FR, BRGM, Montpellier, France

Ladouche B, Maréchal JC, Dörfliger N, Lachassagne P (2006) Système karstique des Cent-Fonts. Simulation de Scénarios d'exploitation et de gestion de la resource. Report BRGM/RP-54865-FR, BRGM, Montpellier, France, 269 p

Ladouche B, Dörfliger N, Maréchal J-C, Lachassagne P, Bakalowicz M, Valarié I, Lenoir P (2007) Hydrodynamic behaviour during pumping test and modelling of the cent fonts karst system. In: Chery L, de Marsily G (eds), IAH Selected papers on hydrogeology, aquifer systems management: Darcy's legacy in a world of impending water shortage, vol 10, pp 303–316

Maréchal JC, Dewandel B, Subrahmanyam K (2004) Use of hydraulic tests at different scales to characterize fracture network properties in the weathered-fractured layer of a hard rock aquifer. Water Resour Res, 40(11)

Maréchal JC, Ladouche B, Dorfliger N, Lachassagne P (2008) Interpretation of pumping tests in a mixed flow karst system. Water Resour Res 44(5)

Ramey HJ, Gringarten AC (1976) Effect of high-volume vertical fractures on geothermal steam well behavior. U.S. Government Printing Office, pp 1759–1762

Renard P, Glenz D, Mejias M (2009) Understanding diagnostic plots for well-test interpretation. Hydrogeol J 17(3):589–600

Application of Methods for Resource and Source Vulnerability Mapping in the Orehek Karst Aquifer, SW Slovenia

Ana Isabel Marín, Nataša Ravbar, Gregor Kovačič, Bartolomé Andreo and Metka Petrič

Abstract In a shallow karst aquifer in SW Slovenia assessment of groundwater and drinking water source vulnerability has been performed using different methods. The small, but well defined, Orehek karst is mainly drained by the no longer exploited Korentan spring. The recharge area of the spring is well karstified, densely wooded, and covered by thin soils. Besides precipitation, the spring is additionally recharged by temporally active sinking streams. To assess the vulnerability, two methods, COP+K and Slovene approach, have been applied. Both methods provide tools for assessing vulnerability in carbonate rocks. They both rank among very sophisticated methods that require vast amounts of data, time, and financial and technical resources. Both methods share the same methodological procedure and consider the same type of information, categorized by the same factors. On the other hand, the scoring, classification and weighting of individual parameters between the methods are different. Consequently, the resulting maps differ significantly. Major differences between the results are identified and discussed.

A. I. Marín (✉) · B. Andreo
Department of Geology, Faculty of Sciences, Centre of Hydrogeology (CEHIUMA), University of Malaga, Malaga, Spain
e-mail: aimarin@uma.es

B. Andreo
e-mail: andreo@uma.es

N. Ravbar · M. Petrič
Karst Research Institute, ZRC SAZU, Titov trg 2, 6230 Postojna, Slovenia
e-mail: natasa.ravbar@zrc-sazu.si

M. Petrič
e-mail: petric@zrc-sazu.si

G. Kovačič
Faculty of Humanities and Science and Research Centre of Koper, University of Primorska, Titov trg 5, 6000 Koper, Slovenia
e-mail: gregor.kovacic@fhs.upr.si

J. Mudry et al. (eds.), *H2Karst Research in Limestone Hydrogeology*, Environmental Earth Sciences, DOI: 10.1007/978-3-319-06139-9_10,

In addition, sensitivity analysis of individual factors have been performed and compared to cross-correlation, autocorrelation and water budget calculations. The results show high dependence of the COP+K method to the parameters characterizing the infiltration conditions (the so-called C factor) and distance to water source. The Slovene approach vulnerability classes are mainly influenced by thickness of the soil cover, presence of morphological features and the temporal hydrological variability that is justified by the performed statistical analysis.

Keywords Vulnerability assessment • Karst aquifer • Water protection • Slovenia

1 Introduction

Karst aquifers are becoming economically more and more important as they contain large amounts of good quality groundwater resources (Ford and Williams 2007). Due to special functioning and behaviour of karst aquifers (high permeability of aquifer systems, concentrated recharge and rapid infiltration of water, fast flow through karst conduits and transport over large distances), karst water sources are particularly vulnerable to contamination. Therefore, they need proper protection and management.

Many national regulations concerning the protection of water sources are primarily based on the distance from the water source or the velocity of the groundwater flow to a spring. Often, the criteria do not consider the special features of water flow in karst, such as the heterogeneity and complexity of recharge conditions, the changes in the velocity and direction of water flow under diverse hydrological conditions, etc. However, almost two decades ago a concept of groundwater vulnerability mapping was recognized as an alternative to conventional regulations regarding the water protection (Vrba and Zaporozec 1994). In some countries, such as Switzerland and Ireland, the concept of vulnerability assessment is successfully used in determining water protection zones and planning land use on karst (Dörfliger and Zwahlen 1998; GPS 1999).

For such assessments, guidelines were elaborated, i.e., the European approach, in the framework of the international COST Action 620 project (Daly et al. 2002; Zwahlen 2004). On these foundations, numerous methods for assessing and mapping the vulnerability of karst waters were developed that took into account differences between individual carbonate aquifer systems, accessibility of data, and economic capabilities. These methods were used and tested at various test sites around the world on a number of occasions. Reviews of some methods have been done by several authors, including Zwahlen (2004), Ravbar (2007), Goldscheider (2010), and others.

Up to now, most complete implementation of the European approach and the most comprehensively designed methods are COP+K (Vías et al. 2006; Andreo et al. 2009) and the Slovene approach (Ravbar and Goldscheider 2007). Although these methods share common methodological procedures and require the same information, a comparison study in a selected test site (Orehek karst) in SW Slovenia has been made. The aim of the study was to distinguish the main differences between the two methods in the resulting resource and source maps. The maps have been evaluated and sensitivity analysis of individual factors has been performed. The obtained results have been evaluated by different statistical analysis of hydrological data.

2 Description of the Test Site

The Orehek karst area is a well defined shallow karst aquifer, situated in SW Slovenia. It covers an area of around 9 km^2. A slightly uplifted ridge reaches up to 725 m in height above the surrounding 545–600 m altitudes. The aquifer consists of an anticline of Cretaceous and Palaeocene limestones (Fig. 1), which is in the SW partially thrusted over Eocene flysch and encircled by it (Gospodarič et al. 1970; Petrič and Šebela 2004). The Orehek karst is well-karstified with numerous dolines and 70 caves registered in the Slovene Cave Cadastre (2010). The carbonate rock in its northern part is covered with a thin layer of rendzina soil, and in the southern part with brown carbonate soil of various depths. The area is influenced by sub-continental climate with average annual precipitation of about 1,670 mm (period 1971–2000). In the same period an average annual runoff was about 900 mm.

The aquifer is additionally recharged by several small sinking rivers on the SW (e.g., Črmelice, Orehovške ponikve) that are not permanent. On the NE margin of the aquifer very low permeable flysch acts as a hydrological barrier. Two temporal springs (Poliček and Mrzla jama) and a permanent Korentan spring drain the aquifer. Tracer test results proved the connection of the southern part of the aquifer and the Orehovške ponikve sinking stream with the Poliček spring at high waters. At low waters, the underground waters of the southern part of the aquifer flow towards the Pivka River situated further to the east (not shown on Fig. 1) (Gospodarič et al. 1970). The second tracer test performed at medium waters proved the connection of the Črmelice sinking rivers with the Korentan spring. The groundwater flow velocity was estimated to be 25 m/h and the tracer recovery at 71 % (Schulte 1994).

The main outflow from the Orehek karst aquifer, the Korentan spring, is a typical karst spring, characterized by rapid, sharp responses to precipitation events. Spring discharges range from a few l/s to about 3 m^3/s with an average of 0.2 m^3/s. Based on the results of tracer tests and the water balance method, the extent of the spring's recharge area has been estimated and delineated to 6 km^2 (including the non-karst part of the catchment: Petrič and Šebela 2004). The results of the water balance calculations infer, that the Orehovške ponikve sinking stream drainage area (marked as "Temporary, not confirmed" on Figs. 2 and 4) does not belong to the Korentan spring catchment.

The results of daily and hourly time series analysis show low memory effect for discharge and high memory effect for electrical conductivity and

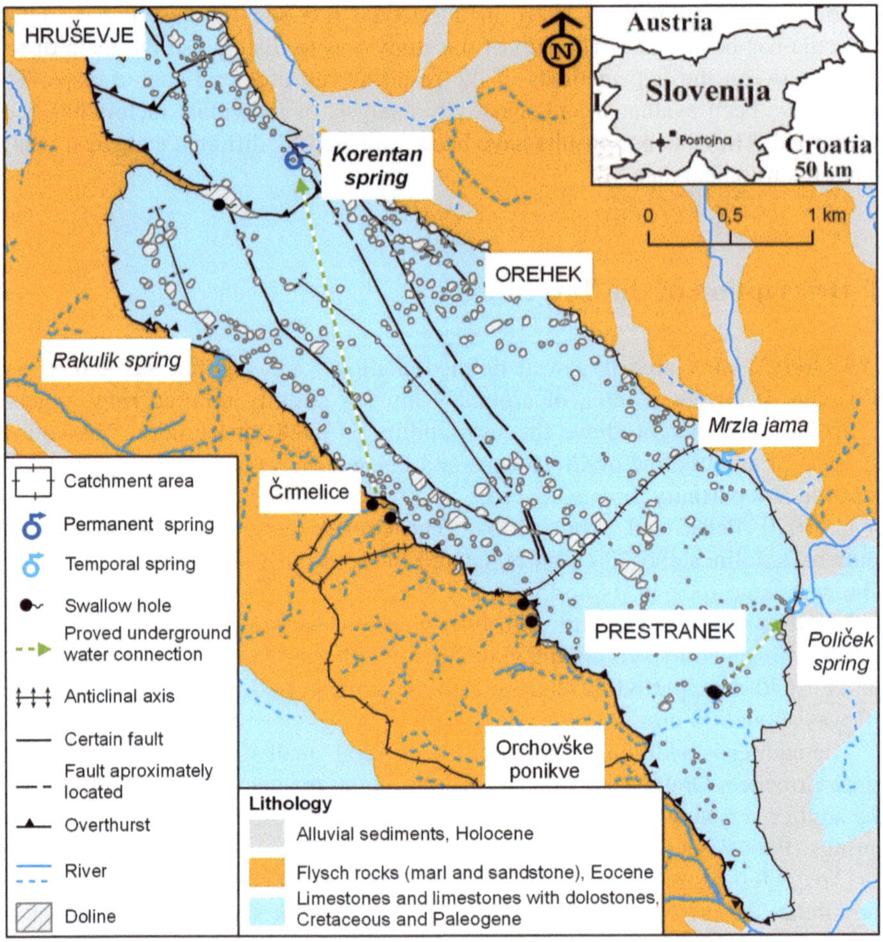

Fig. 1 Hydrogeological map of the Orehek karst (after Gospodarič et al. 1970)

temperature of the Korentan spring. The storage capacity of the Orehek karst aquifer is therefore small, meaning that the spring reacts instantly to the precipitation events in the catchment, which is typical for small and well karstified karst systems (Kovačič 2010). Small variations in the values and high memory effect of electric conductivity and temperature of the spring show that the autogenic recharge is of much greater importance for the spring as the recharge from the Črmelice sinking river.

Korentan has been used for local drinking water supply in the period 1955–1972. A few years ago its reactivation as a reserve source has been discussed and a source vulnerability map using the EPIK method has been prepared in order to determine the water protection zones (Fig. 2). The map shows predominance of Moderate to High vulnerability of the aquifer. Very high vulnerability has been

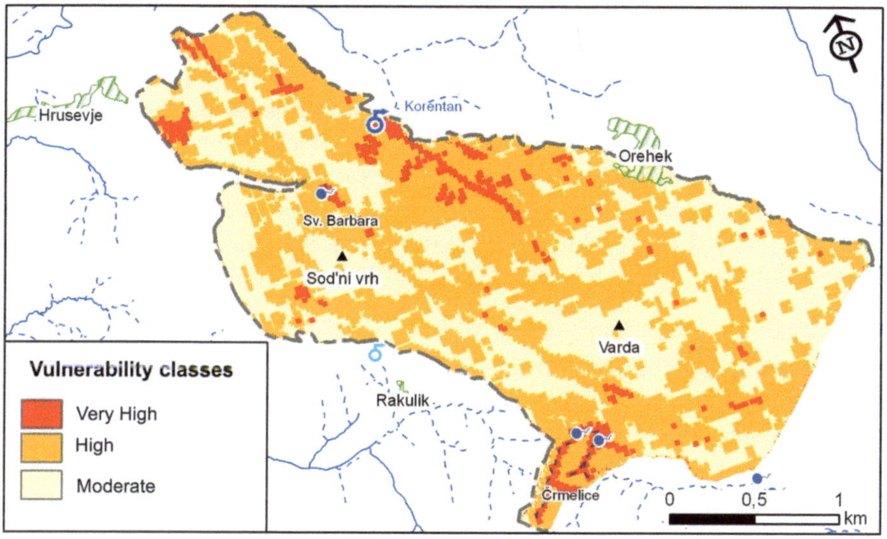

Fig. 2 Vulnerability map obtained with the EPIK method (Petrič and Šebela 2004)

assigned to sinkhole drainage basins and to areas with well-developed epikarst (Petrič and Šebela 2004; Kovačič and Petrič 2007). Unfortunately, until now no protection zones and regimes have been implemented and no further effort for the spring's re-utilization has been made.

3 Methodology

In this study, two methods for assessing groundwater vulnerability, the COP+K (Vías et al. 2006; Andreo et al. 2009) and the Slovene approach (Ravbar and Goldscheider 2007) have been applied in the Orehek karst. Both methods take into account the specific properties of karst. The vulnerability assessment includes the geological, hydrological, and hydrogeological characteristics of a karst system and the precipitation regime. Relative to the purpose, two types of vulnerability assessment are available: for resource and for source (Daly et al. 2002). Assessment of resource vulnerability includes parameters that control the flow of infiltrated water from the surface all the way to the water table. Here, relevant factors include the permeability and thickness of the soil and rock composing the unsaturated zone (named O factor) and the concentration of runoff as influenced by topography, the karst features, the vegetation cover (C factor), and the distribution and intensity of precipitation (P factor). The additional K factor, which considers the characteristics of water flow in the saturated zone, makes it possible to assess the vulnerability of a water source.

The COP method (Vías et al. 2006) was primarily designed for resource vulnerability assessment, but was later adapted for source vulnerability assessment and named COP+K (Andreo et al. 2009). The Slovene approach (Ravbar and Goldscheider 2007) is based on the COP method for assessing resource vulnerability; therefore, some of the factors are differently scored or classified. In addition, the Slovene approach has been extended for source vulnerability mapping. It also offers the possibility of considering the temporal variability of hydrological conditions and the linked protection of surface waters and groundwater. The method additionally includes guidelines for hazard and risk assessment, and considers the value or importance of a groundwater source or resource.

In order to analyze and compare the resulting maps, the sensitivity analysis techniques have been utilized. This technique provides valuable information on the influence of each parameter on the final index. The maps have been fragmented to the "unique condition subareas" (UCS). A UCS contain one or more zones (consisting of pixels) where a unique combination of different parameters compose the final vulnerability index (Gogu and Dassarges 2000). Then the sensitivity can be evaluated as:

$$S_{xi} = \frac{V_i}{N} - \frac{V_{xi}}{n} \tag{10.1}$$

where S_{xi} is the sensitivity (for the ith UCS) associated with the removal of one map (of the X factor), V_i is the vulnerability index original on the ith subarea, V_{xi} is the vulnerability index of the ith subarea calculated without considering the X factor, N is the number of maps used in primary suitability (3) and n is the number of maps used in perturbed suitability (2).

The sensitivity of each parameter for the whole catchment area is:

$$S = \frac{\sum_{i=1}^{i=m} (S_{xi} \cdot \Delta_{xi})}{m} \tag{10.2}$$

where S is the sensitivity associated with the removal of one map from the whole catchment area, S_{xi} is the sensitivity (for the ith UCS) associated with the removal of one map (of the X factor), Δ_{xi} is the percentage of surface area of each UCS and m is the number of UCS.

As the values of the O factor vary from 1 to 15, while those of the C and P factors vary from 0 to 1, it is necessary to divide the O value by the maximum value possible (15). Thus, all the variables have the same range (0–1) and the distortion produced in the sensitivity analysis results by the differences between the values of the factors is eliminated.

This technique was applied, amongst others, by Gogu and Dassargues (2000) for the EPIK method, by Pételet-Giraud et al. (2000) for the RISKE method and by Marín et al. (2012) for the PaPRIKa and COP methods.

Cross-correlation analysis of daily precipitation-discharge data, autocorrelation analysis of temperature and electric conductivity of the Korentan spring and water budget calculations of the aquifer have additionally been made to better characterize the aquifer system.

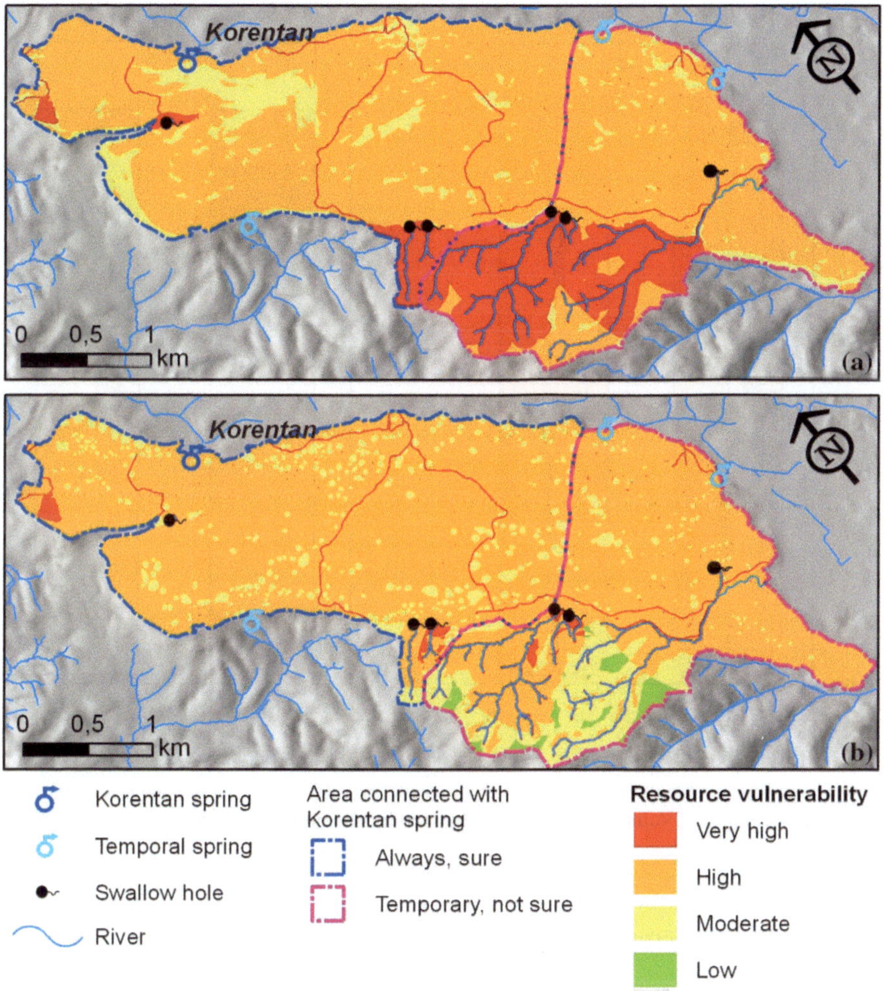

Fig. 3 Resource vulnerability maps obtained with the COP method (**a**) and the Slovene approach (**b**)

4 Vulnerability Maps: Results and Comparison

Figures 3 and 4 show the resource vulnerability maps and the percentages of vulnerability classes. According to both methods, the karst aquifer is generally characterized by High vulnerability with some patches of Moderate vulnerability. This is using the COP method influenced merely by the land use, density of vegetation cover and slope inclination, while using the Slovene approach by the thickness of the soil cover, location of the morphological features and the temporal variability of the hydrological conditions.

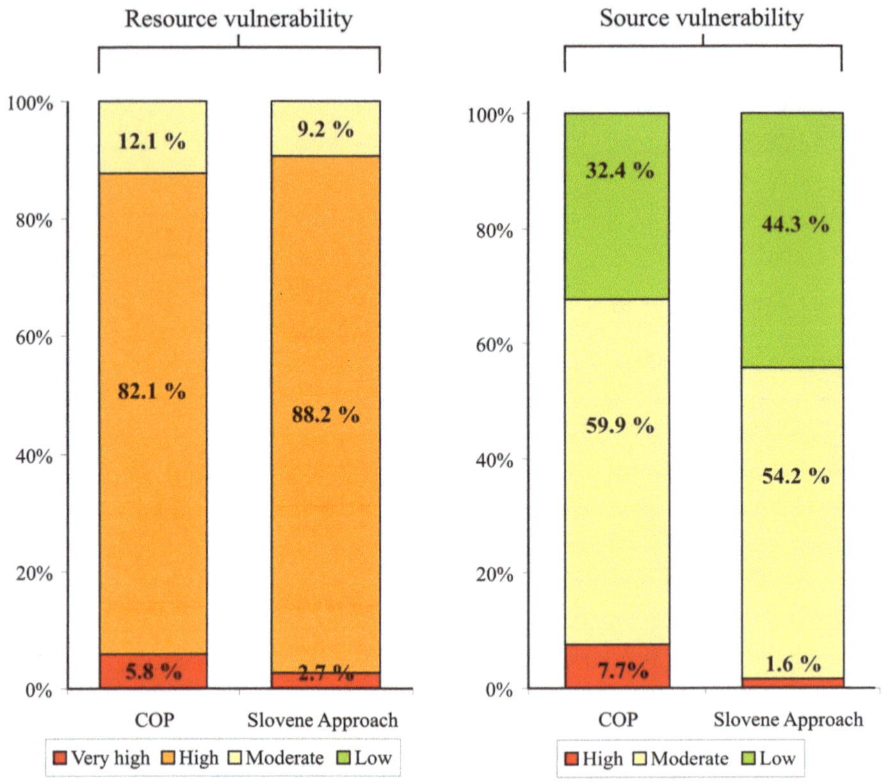

Fig. 4 Percentages of vulnerability classes obtained by the applied methods

Table 1 Average sensitivities of parameters for the whole aquifer

	UCS	C	O	P
COP method	45	0.17	0.25	0.04
Slovene approach	53	0.11	0.29	0.03

The main difference between the vulnerability methods concerns the catchment areas of the sinking streams. These are characterized by significantly lower vulnerability by the Slovene approach due to consideration of temporal variability of hydrologic conditions and lower score of the variable distance to swallow holes.

The results of the overall sensitivity analysis show, that both vulnerability assessments are highly sensitive to the O factor, less sensitive to the C factor and almost insensitive to the P factor (Table 1). The insignificantly small influence of the P factor to the final degree of vulnerability using the COP method has been previously also reported by Vías et al. (2010).

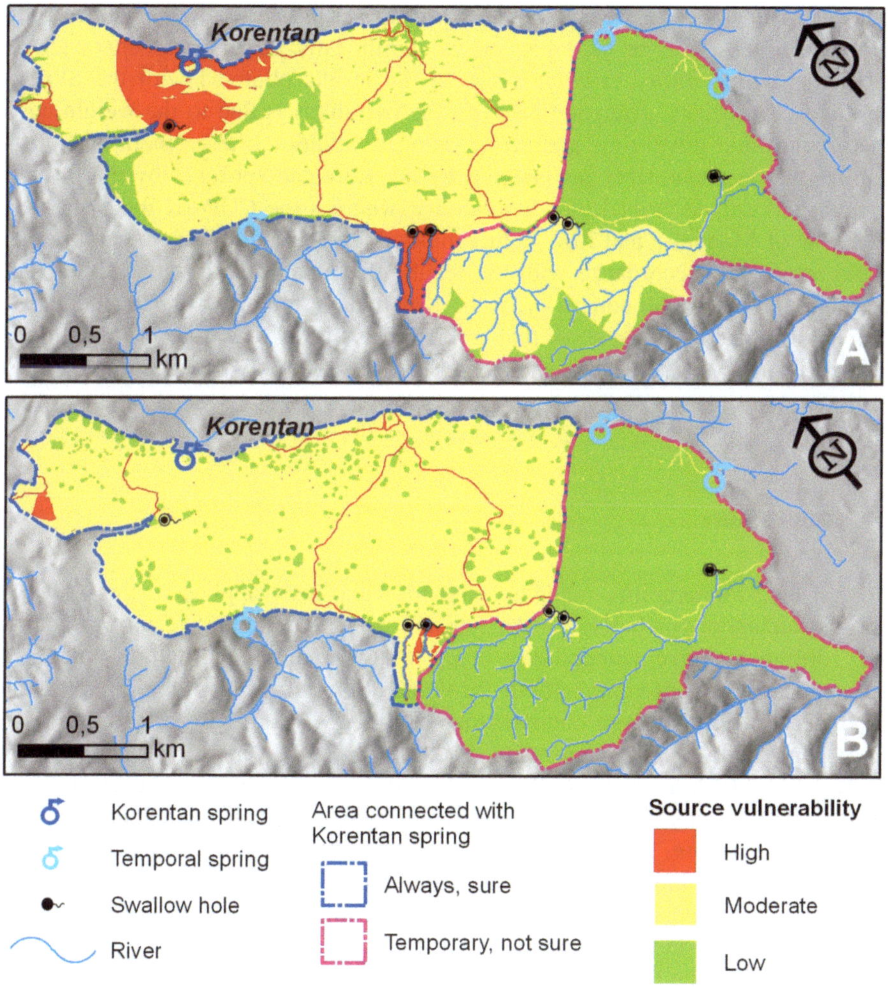

Fig. 5 Korentan source vulnerability maps obtained with the COP method (**a**) and the Slovene approach (**b**)

Figures 4 and 5 show the Korentan source vulnerability maps and the percentages of vulnerability classes. Evaluation of source vulnerability mainly depends on data of predominant groundwater flow directions and velocities.

These resulting maps are consistent with the results of the daily precipitation-discharge cross-correlation analysis that point to high vulnerability of the Orehek karst aquifer as well. The results proved fast reaction of the Korentan karst spring discharges to precipitation events and low storage capacity of the system, which is characterised by shallow vadose zone.

Both, the COP+K and the Slovene approach methods clearly differentiate the direct recharge area from the southern sector of the aquifer, the connection of which has not been proven with the Korentan spring. In this area, vulnerability is considered Low. The COP+K shows significantly larger areas characterized by higher vulnerability in comparison to the results obtained with the Slovene approach. A distinct difference between the two maps is due to the weighting of the groundwater travel time and the K factor. In addition, the southern part of the aquifer that is presumably not drained by the Korentan spring is by the Slovene approach classified as of Low vulnerability, whereas the COP+K method classifies it as of Moderate and Low vulnerability. Nonetheless, lower vulnerability of these areas is in accordance with the results of autocorrelation analysis of temperature and electric conductivity of the Korentan spring and water budget calculations of the aquifer that show that sinking rivers have very little or insignificant influence on spring water. The Korentan spring is mainly recharged by diffuse infiltration.

5 Conclusion

So far three vulnerability mapping methods have been applied in the Korentan spring catchment (including EPIK; Kovačič and Petrič 2007) resulting in three very different source vulnerability maps. Many studies have demonstrated that the application of different methods for vulnerability mapping at the same test site produced different or even contradictory results despite the use of the same data (e.g., Vías et al. 2005; Neukum and Hötzl 2007; Ravbar and Goldscheider 2009; Marín et al. 2012, and others). The present study showed that it is even so in the case when methods are based on the same methodological procedure, the same type of information, categorized by the same factors. The scoring, classification and weighting of individual parameters plays the most important role in assessing vulnerability degree. However, unless the results are verified, it is not certain which method produces the most reliable results.

In this study, the COP+K method and Slovene approach that are very similar methods have been further analyzed by sensitivity, cross-correlation and autocorrelation analysis. The results show that the COP+K vulnerability maps are merely influenced by the land use, karst features, vegetation density and slope inclination data that compose the so-called C factor, and by the distance to the water source. The Slovene approach vulnerability maps are, on the other hand, influenced by thickness of the soil cover, morphological features and temporal vulnerability. The performed correlation analysis showed high vulnerability of the aquifer, its fast reaction to precipitation events and low storage capabilities, and justifies the introduction of temporal hydrological variability into the vulnerability mapping.

In the future, various conventional hydrogeological methods and techniques (hydrograph and chemograph analyses, tracer tests at different hydrological conditions), and previous studies of hydrodynamic response of the Korentan spring (Jemcov and Petrič 2009; Petrič and Šebela 2004) should be used to evaluate the obtained vulnerability maps and to verify the used methods in detail. As only reliable and objective interpretation of vulnerability indices can serve as a basis for the alternative protection zoning in karst, these should be as reliable as possible.

Acknowledgments This manuscript is a contribution to projects CGL2008-06158 BTE of the Spanish Ministry of Science and Higher Education and IGCP 598 of UNESCO, and to Research Group RNM-308 funded by the Regional Government of Andalusia (Spain).

References

Andreo B, Ravbar N, Vías JM (2009) Source vulnerability mapping in carbonate (karst) aquifers by extension of the COP method: application to pilot sites. Hydrogeol J 17(3):749–758

Daly D, Dassargues A, Drew D, Dunne S, Goldscheider N, Neale N, Popescu C, Zwahlen F (2002) Main concepts of the European approach for (karst) groundwater vulnerability assessment and mapping. Hydrogeol J 10(2):340–345

Dörfliger N, Zwahlen F (1998) Practical guide: groundwater vulnerability mapping in karstic regions (EPIK). Swiss Agency for the Environment, Forests and Landscape, Bern

Ford DC, Williams PW (2007) Karst hydrogeology and geomorphology. Wiley, London

Gogu RC, Dassargues A (2000) Sensitivity analysis for the EPIK method of vulnerability assessment in a small karstic aquifer, southern Belgium. Hydrogeol J 8:337–345

Goldscheider N (2010) Delineation of spring protection zones. In: Kresic N, Stevanovic Z (eds) Groundwater hydrology of springs. Engineering, theory, management, and sustainability. Butterworth-Heinemann, Oxford

Gospodarič R, Habe F, Habic P (1970) Orehovški kras in izvir Korentana (the karst of Orehek and the source of the Korentan). Acta Carsologica 5:95–108

GPS: Groundwater Protection Schemes (1999) Dublin, Department of the Environment and Local Government, Environmental Protection Agency and Geological Survey of Ireland

Jemcov I, Petrič M (2009) Measured precipitation vs. Effective infiltration and their influence on the assessment of karst system based on results of the time series analysis. J Hydrol 379:304–314

Kovačič G (2010) Hydrogeological study of the Malenščica karst spring (SW Slovenia) by means of a time series analysis. Acta Carsologica 39(2):201–215

Kovačič G, Petrič M (2007) Karst aquifer intrinsic vulnerability mapping in the Orehek area (SW Slovenia) using the EPIK method. In: Witkowski AJ (ed) Groundwater vulnerability assessment and mapping: selected papers from the groundwater vulnerability assessment and mapping international conference, Ustroń, Poland, 2004. Taylor and Francis, London

Marín AI, Dörfliger N, Andreo B (2012) Comparative application of two methods (COP and PaPRIKa) for groundwater vulnerability mapping in Mediterranean karst aquifers (France and Spain). Environ Earth Sci 65(8):2407–2421

Neukum C, Hötzl H (2007) Standardization of vulnerability maps. Environ Geol 51:689–694

Petrič M, Šebela S (2004) Vulnerability mapping in the recharge area of the Korentan spring, Slovenia. Acta Carsologica 33(2):151–168

Pételet-Giraud E, Dörfliger N, Crochet P (2000) RISKE: Méthode d'évaluation multicritère de la vulnérabilité des aquifères karstiques. Application aux systèmes des Fontanilles et Cent-Fonts (Hérault, Sud de la France). Hydrogéologie 4:71–88

Slovene Cadastre of Caves (2010) Karst Research Institute at ZRC SAZU

Schulte U (1994) Geologische und Hydrogeologische Untersuchungen im Karst von Orehek (Slowenien). Diplomarbeit, Universität Karlsruhe, Deutschland

Ravbar N (2007) The protection of karst waters—a comprehensive Slovene approach to vulnerability and contamination risk mapping. ZRC Publishing, Karst Research Institute at ZRC SAZU 6: Carsologica, Postojna, Ljubljana

Ravbar N, Goldscheider N (2007) Proposed methodology of vulnerability and contamination risk mapping for the protection of karst aquifers in Slovenia. Acta Carsologica 36(3):461–475

Ravbar N, Goldscheider N (2009) Comparative application of four methods of groundwater vulnerability mapping in a Slovene karst catchment. Hydrogeol J 17:725–733

Vías JM, Perles MJ, Andreo B, Carrasco F (2005) A comparative study of four schemes for groundwater vulnerability mapping in a diffuse flow carbonate aquifer under Mediterranean climatic conditions. Environ Geol 47:586–595

Vías JM, Andreo B, Perles MJ, Carrasco F, Vadillo I (2006) Proposed method for groundwater vulnerability mapping in carbonate (karstic) aquifers: the COP method: application in two pilot sites in southern Spain. Hydrogeol J 14(6):912–925

Vías JM, Andreo B, Ravbar N, Hötzl H (2010) Mapping the vulnerability of groundwater to the contamination of four carbonate aquifers in Europe. J Environ Manage 91(7):1500–1510

Vrba J, Zaporozec A (1994) Guidebook on mapping groundwater vulnerability. IAH, International Contributions to Hydrogeology, 16, Heise, Hannover

Zwahlen F (ed) (2004) Vulnerability and risk mapping for the protection of carbonate (karst) aquifers. Final report of COST Action 620. European Commission, Directorate-General XII Science, Research and Development, Brussels

Study of Subterranean Floods in Oceanic Subpolar Karst of Madre de Dios Archipelago (Patagonia, Chile)

Laurent Morel, Stéphane Jaillet, Richard Maire, and Members of Ultima Patagonia

Abstract Madre de Dios archipelago is located, at 50°30′ S, on the Pacific front. The karst areas in the reef limestone of Upper Paleozoic of Chilean Patagonia have long remained unexplored because of their remoteness, difficult access and very inhospitable cold, wet and windy climate. The annual rainfall is 7–8 m/year^{-1} and the average wind speed reaches 70 kmph^{-1} almost unidirectional (W to NW), and involve strong floods at a high rate. To study the flood dynamics, several underground sites have been instrumented in 2008 and recovered in 2010.

Keywords Flood • Karst • Cave • Instrumentation • Patagonia • Chile

1 Introduction

Madre de Dios archipelago is located at 50°30′ S, on the Pacific front, on the isotherm +7 °C, at the northern limit of the subpolar isothermic climate (Maire et al. 1999; Fig. 1). The karst areas of Chilean Patagonia have long remained unexplored because of their remoteness, difficult access and very inhospitable cold, wet and windy climate.

L. Morel (✉)
Laboratoire Ampère, Université Lyon 1 et Ecole Centrale de Lyon, Lyon, France
e-mail: laurent.morel@univ-lyon1.fr

S. Jaillet
Laboratoire EDYTEM, Université de Savoie CNRS, Campus Scientifique, 73376 Le Bourget du Lac, France

R. Maire
Programme ANR Climanthrope, Laboratoire ADES, Université de Bordeaux 3—CNRS, Maison des Suds, 33607 Pessac Cedex, France

L. Morel · S. Jaillet · R. Maire · Members of Ultima Patagonia
Association CENTRE—TERRE, 25 rue Louis de Broglie, 31100 Toulouse, France

J. Mudry et al. (eds.), *H2Karst Research in Limestone Hydrogeology*, Environmental Earth Sciences, DOI: 10.1007/978-3-319-06139-9_11,

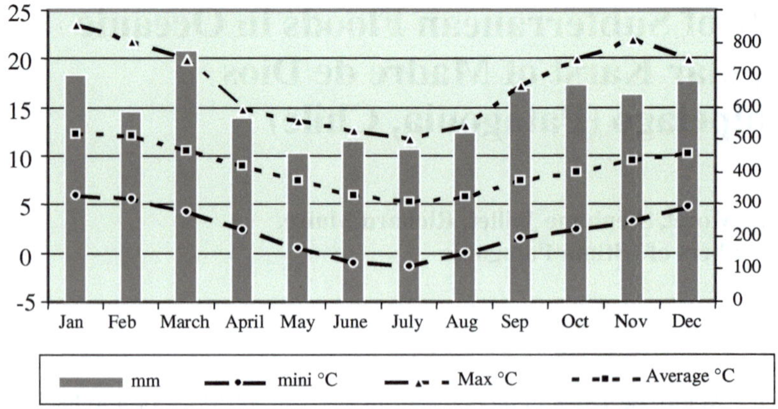

Fig. 1 Average of rainfall and temperatures in Guarello during the 2004–2007 period

Fig. 2 Location map of Madre de Dios archipelago

The average wind speed reaches 70 kmph^{-1} and the winds can regularly exceed 100 kmph^{-1} (Fig. 2). The annual thermal amplitude is weak, about 5–6 °C. The daytime temperatures are rarely negative at sea level. Annual rainfall is 7–8 m/year^{-1}; it is very well-balanced, with average precipitation evenly distributed throughout the year.

In protected places, the vegetation is characterized by the primitive magellanic forest with the genus Nothofagus inherited of Gondwana. The southern edge of Madre de Dios was almost completely covered by ice during the last cold period, except for some limestone nunataks. The Permian and Carboniferous limestones (Tarlton limestones) are located between volcano-sedimentary formations of Upper Paleozoic on South and West (Duque de York formation) and the Mesozoic granites of the Patagonian Batholith on the east. These carbonates, with many dikes, correspond to corallian paleoreefs, part of an accretionary prism of the Gondwana paleocontinent. Recent Kr–Ar dating of biotite indicates 133–140 Ma for the intrusions in limestones related to the magmatic activity of an ancient volcanic arc (Sugden et al. 2005; Maire et al. 2009a). Some residual fragments of the old oceanic floor (Denaro Complex) were observed.

The first reconnaissance of the French association Centre-Terre was made in 1995 by a four-caver team aboard a tiny fishing boat and allowed a brief incursion on Diego de Almagro at 51°30′ S (Pernette et al. 2009). The aim was to verify the presence of karst and possible cave systems on some of the isolated islands. Four other expeditions took place in 1997, 2000, 2006 and 2008. The discovery of Kawésqar remains (burial sites) and paintings, the giant dissolution runnels, the "rock comets" (wind-oriented solution features), the Whale Cave with many whale bones located between +6 and +37 m high (Maire et al. 2009a), and many other discoveries confirmed the important archaeological and geomorphological potential. This exceptional karst, called the "Marble Glaciers", is determined by the absence of frost below 500 m a.s.l., the huge rainfall and the strong winds coming from the Pacific Ocean (Jaillet et al. 2009, 2010, 2011; Maire et al. 2009b).

To study the flood dynamics and their frequency, different underground sites have been instrumented in 2008 and the data recovered in 2010. During heavy rainfall most of the caves are partially flooded. In order to study and to understand these systems, three kinds of different underground networks were instrumented with "Luirographes", special data loggers measuring the pressure and the temperature of water (Morel et al. 2009).

In 2008, four sites have been studied: (1) North Tarlton for exokarst runnels; (2) Mask Sinkhole (Guarello), an underground river fed by three lakes at the contact sandstone-limestone; (3) Plein Cintre cave (Madre de Dios), a young karst spring situated on Soplador hillside; and (4) Kawtcho Sinkhole located at the bottom of the valley, in the extension of the Soplador fjord. It collects part of this valley's water with an average flow of a few hundred litres per second.

In these sites the water level (function of water flow) and temperature have been recorded during two years with a time step of three minutes. These sites present different dynamics, with an underground river fed by exokarstic lakes and rivers. The caves develop in contact with sandstone and limestone. The rainfall is very high and involves strong floods. The results will provide knowledge about the rising time, the number and intensity of floods depending on studied sites. The accessibility conditions are an important parameter. All of them can be easily reached in one to four hours walking distance.

2 Data Logger: The Luirographe

The Luirographe is an apparatus to measure and to record the fluctuation of water level (discharge); it has been used for the first time in the Cave of La Luire, Vercors, in the French Alps (Morel 2010). This instrumentation is based on the principle of

Fig. 3 Typical solution runnels with steps in Madre de Dios

water absolute pressure at a specific point. It can continually record for several years with a frequency of acquisition varying from one second to several minutes.

Friction losses: Any type of moving fluid has energy losses depending on the state of the surface (and hence of friction) against the walls. These losses are related to a reduction of pressure (or open water level) and are oriented to the direction of water flow (Morel and Lismonde 2010). In addition, the friction losses increase as a function of the water velocity. These friction losses depend on the wall roughness, but also of obstacles (flow disturbances) such as bends, boulders and any change of the conduit section. The unit is the pressure and does not depend on ambient pressure. This water level is a function of the water flow collected by the cave. Without flow measurement, it is impossible to link the water level with flow.

3 Presentation of the Studied Site

3.1 Exokarst Runnels n°1

The exokarstic runnels are a typical dissolution feature of Madre de Dios (Fig. 3). These drainage forms represent hydrologic networks in miniature. All the major parameters of terrestrial karst erosion are there: fluvial erosion, meander formation, regressive erosion. Runnel floods have been observed at nearly every rainfall.

The rate of surface dissolution is estimated at more than 100 mm/millennium from glacial erratics, exhumed dykes or historical paintings, the most important known in the world. The experimental site (2008–2010) has been achieved to understand the karstic dissolution on limestone in subpolar oceanic setting (Jaillet et al. 2010, 2011).

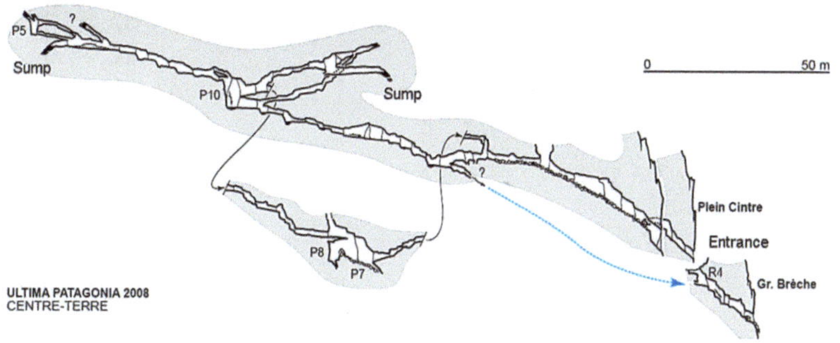

Fig. 4 Cross section of Plein Cintre cave

3.2 Karst Spring of Plein Cintre, Site n°2

The cave of Plein Cintre is a young karst spring situated on the left side of Soplador valley. The entrance is a large porch and the cave is a labyrinth galleries system. This cave is crossed by an underground river cut by several waterfalls and ending with a sump (Fig. 4). In the upper galleries some parietal clayey deposits can be observed and show oldest flooded galleries. Remnants of fluvio-glacial deposits (pebbles, sands, silts) are visible in some places. The Luirographe has been installed outside, just near the exit, upstream of a threshold.

The recordings indicate the underground river flow is similar to the exokarstic runnel flow. Floods have been observed at nearly every rainfall. The increasing of water level does not exceed 30 cm. The catchment of about some km^2 is constituted of open cracks and barren karren, and the cave develops very close to the surface.

3.3 Kawtcho Sinkhole, Site n°3

The Kawtcho sinkhole is located in the valley of Seno Soplador at the contact of limestone and sandstone. A river with a flow rate of several tens of l/s enters in the sinkhole. During high waters, the water level can reach up to +25 m; at that moment the cave is flooded completely with a lake in the entrance. The sinkhole drains a part of the water of the Soplador valley. The mean temperature water is 8 °C, and it is colored red brown by tannin of peatbogs. The cave ends with two siphons (±0 m a.s.l.) and the resurgence is probably located under the sea level in the Seno Soplador (Fig. 5).

3.4 Grotte du Masque, Site n°4

The "grotte du Masque" is located at the bottom of a large depression south of Guarello, at the contact with limestones and sandstones. Two rivers disappear inside two different sinkholes separated by a hundred meters. A dry gallery of 4 m

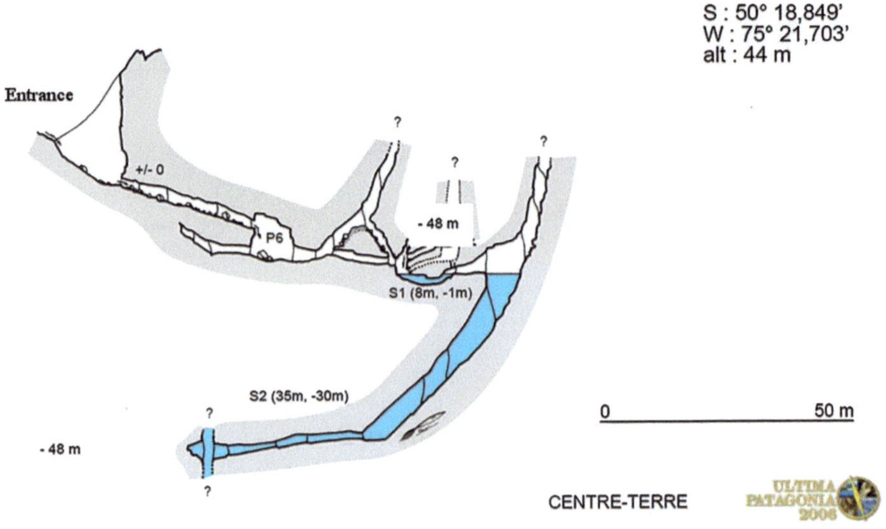

Fig. 5 Cross section of the Kawtcho sinkhole

in diameter, with sediments on the floor, is reached by an upper entrance. Two rivers join each other at 60 m deep and some hundred meters before the terminal siphon (Fig. 7). Outside, the first river is fed with a lake that is fed by another one; this river goes to the *Masque* sinkhole. The *Plume* Sinkhole is fed with the second river that is fed with a third lake (Fig. 6). The Magellanic forest and peatbogs cover a large proportion of catchment estimated to 3 km^2. This configuration has a direct impact on the hydrogeological dynamic, as will be discussed below.

4 Study of Recordings on the Different Sites

The rain falls discontinuously; however, the water flow is smoothed and depends on the studied site. Three parts could be observed on the flood hydrograms: rise and fall times, short recession curve. The recession curve, the rise curve and the fall times are different according to the site. The signal is naturally filtered by the cave. The characteristics of this filter and the time constant are the signature of the karst system.

A hydrogram of Kawtcho sinkhole and Plein-Cintre spring is presented in Fig. 8. The pressure sensor records the atmospheric and water pressure. Therefore, for Plein Cintre the water level is nearly the same value; however, the atmospheric pressure fluctuates more slowly than the water pressure. For example, a reduction of pressure is observed the February 12th and an increase of pressure just after. At midday a flood is observed, but not in the Kawtcho sinkhole. Just before the 12th, a very short reaction time was observed between the two caves of less than one

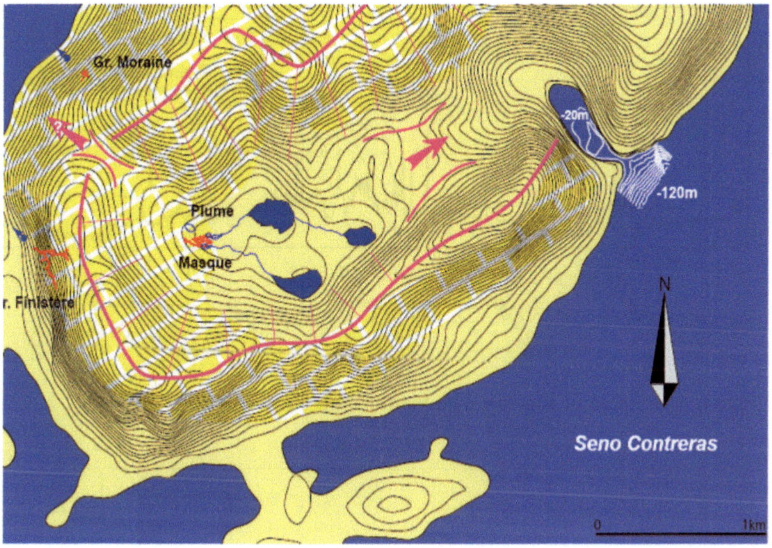

Fig. 6 Situation of the three lakes system and of the Masque cave, Guarello Island

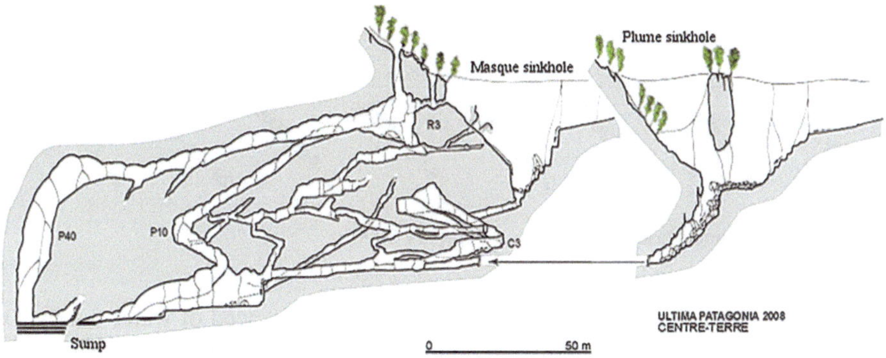

Fig. 7 Cross section of the Masque Cave

hour. The reaction time is a quarter of an hour for the Plein-Cintre from the beginning of the rainfall. Reaction time is confirmed by many floods.

The flood hydrogams of Kawtcho and Masque are presented in Fig. 9. Here again, the curves of Masque cave are smoothed compared to Kawtcho, for example, for the floods of February 24th. The time reaction between the two caves is about one or two hours depending on the floods.

The sorted water level of Kawtcho is presented in Fig. 10 during a two-year period. For more than half the year the cave has at least a few meters of water level. This curve is monotonic and decreasing because of the simple organization of the pipe (tube). The same curve type is found for Masque cave.

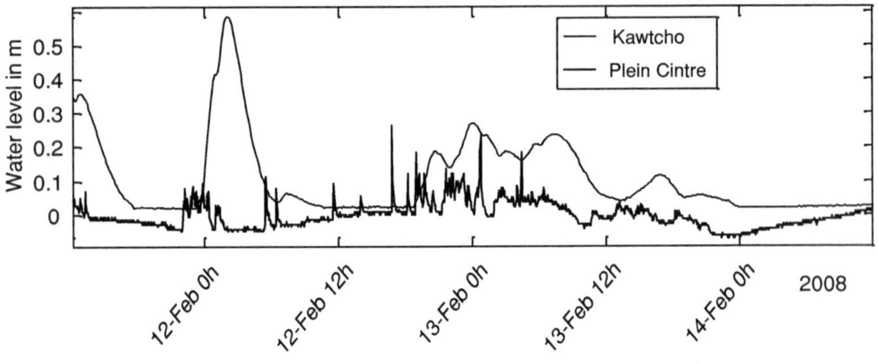

Fig. 8 Floods hydrogram of Kawtcho and Plein-Cintre, February 2008

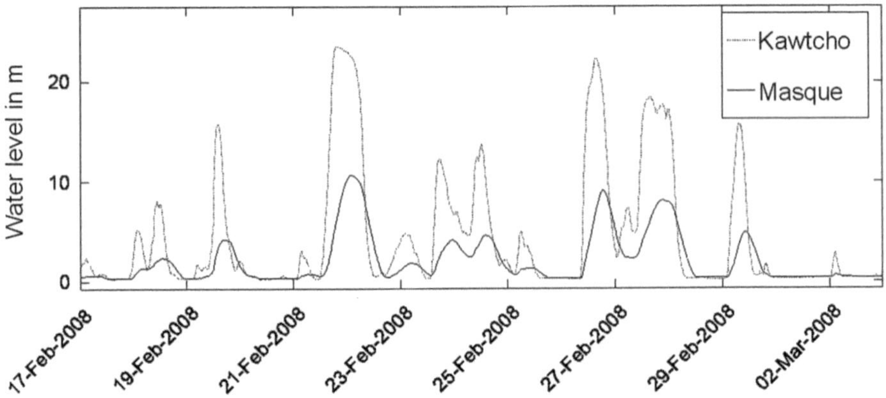

Fig. 9 Kawtcho and Masque hydrograms, February 2008

Another characteristic of these three networks is the speed of water floods. On the hydrograms the flood curves are smoothed and depend on the type of cave. For Plein-Cintre resurgence the time reaction to a rainfall is much quicker than for Kawtcho and the Masque cave. To compare these results, the hydrograms are standardized, that is to say the maximum water level recorded is brought to 100 % for each cave. In other words, the maximum flood "measures" 100 %. The speeds of rising are calculated on these standardized curves (Fig. 11). The unit of these speeds is in percent per hour. For the Plein-Cintre the maximum speed recorded is 65 % per hour; for the Kawtcho it is half and for the Masque six times less.

The ranking of importance of floods is presented in Fig. 12. The water levels are standardized in the aim to be compared (the maximum flood is brought to 100 %). The Masque curve is smoothed for large floods. This graphic represents one of signatures of these networks.

It is possible to determine hydrological characteristics of these networks if the spring discharge is known. However, only a few emergences above sea level have

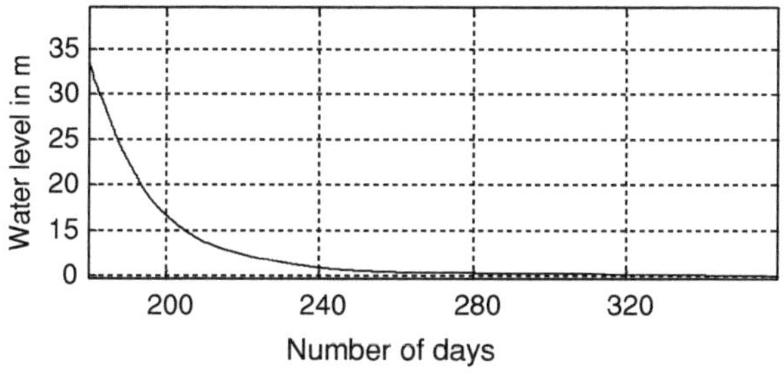

Fig. 10 Kawtcho sorted water level, February 2008

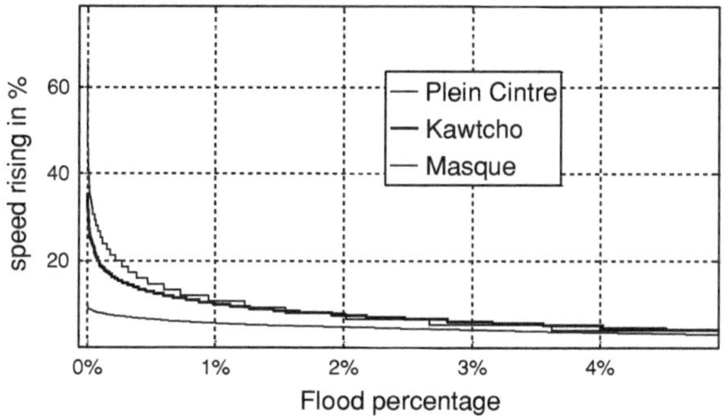

Fig. 11 Speed of rising of standardized curve for the three sites

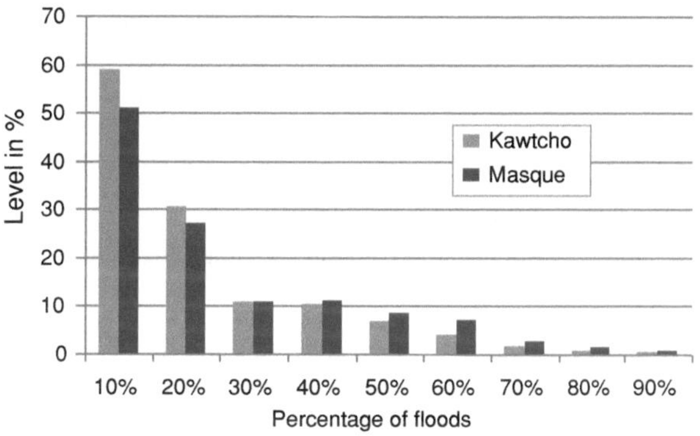

Fig. 12 Ranking of importance of floods

been discovered on Madre de Dios archipelago while other springs are submarine. Most of the karstic systems have been formed during the last ice ages when the ocean level was lower than now.

More examples could be given for other caves. The flash flood that occurred in February 2000 in a sinkhole cave ("Perte du Futur") is remarkable for its suddenness. The response time was only fifteen minutes between the exokarstic flow into the runnels and the water wall running in a steep gallery located 150 m below the surface. There is no reserve of water in the epikarst zone because of the open fractures as shown in the hydrograms of Plein-Cintre. The lack of an epikarstic aquifer is proved by the flow regime. Without rainfall, the subterranean rivers are fed only by water stockage situated in the impervious sandstones. For example, this water storage feeds the Masque sinkhole. The only reserves of water are situated in the phreatic zone; however, it is difficult to estimate the dynamic volume and the permanent volume without true recession curve.

5 Conclusion

Four sites have been studied: a resurgence, an exokarstic runnel, a system of lakes and a sinkhole. Several types of subterranean floods are presented in these four examples. Many floods have been recorded because of high rainfall spread throughout the year. These hyperhumid, karsts situated in the Roaring Fifties, in the subpolar oceanic zone, have allowed a better understanding of this very original underground hydrology. An important research work must be continued.

Acknowledgments This work was assisted by the financial support of ANR Climanthrope program and human resources and logistics support of Utima Patagonia expeditions 2006, 2008 and 2010. http://www.centre-terre.fr

References

Jaillet S, Morel L, Maire R, Malet E, Hoblea F, Ployon E (2011) Instrumentation et dissolution superficielle des karsts subpolaires océaniques de l'Archipel de Madre de Dios (Patagonie, Chili). 9ème colloque d'hydrogéologie en pays calcaire sur les aquifères, H2KARST Besançon

Jaillet S, Morel L, Maire R, Malet E, Hoblea F, Ployon E (2010) Taux record de dissolution superficielle des karsts de Patagonie: instrumentation et quantification sur le bassin versant expérimental de Tarlton. Réunion des Sciences de la Terre, Bordeaux

Jaillet S, Maire R, Brehier F, Despain J, Morel L (2009) Englacement, eustatisme et réajustements karstiques de la bordure sud de l'archipel de Madre de Dios (Patagonie, Province última Esperanza, Chili). Kartologia 2008, pp 1–24

Maire R, l'Equipe Ultima Esperanza (1999) Les "glaciers de marbre" de Patagonie, Chili. Un karst subpolaire océanique de la zone australe. Karstologia 33:25–40

Maire R, Tourte B, Jaillet S, Despain J, Lans B, Brehier F, Fage L-H, Morel L (2009a) Geomorphic and archaeological features of coastal caves in Madre de Dios archipelago (Patagonia, Chile). In: Proceeding of "Karst Horizons" international congress, vol 3, UIS, Kerrville, Texas, pp 516–521

Maire R, Jaillet S, Hobléa F (2009b) Karren in Patagonia, a natural laboratory for hydroaeolian dissolution. In: Ginés A et al (eds) Karst rock features, Karren sculpturing. Carsologica 9, ZRC Publishing, Postojna-Ljubljana, pp 329–348

Morel L (2010) Le Luirographe: un outil pour l'étude des systèmes karstiques. Application au réseau de la Luire (Vercors, France). Karstologia 17:88–93

Morel L, Lismonde B (2010) Crues et mises en charge. In: Grottes et Karsts de France, sous la dir. de Ph. Audra, Karstologia Mémoire vol 19, pp 28–29

Morel L, Jaillet S, Perroux A-S, Maire R, Delannoy J-J, Perrette Y, Lignier V, Malet E, Ultima Patagonia (2009) Karst instrumentation to study site effect examples in Choranche cave (France) and Madre de Dios archipelago (Chile). In: Proceeding of "Karst Horizons" international congress, vol 1. UIS, Kerrville, Texas, pp 591–596

Pernette J-F, Tourte B, Maire R (2009) The Centre-Terre expeditions to patagonian karst islands: a historic overview. In: Proceeding of "Karst Horizons" international congress, vol 3., UIS, Kerrville, Texas, pp 1860–1865

Sugden D, Bentley M, Fogwill C, Hulton N, Mcculloch R, Purves R (2005) Lateglacial events in southernmost South America: a blend of "Northern" and "Southern" hemispheric climatic signals. Geografiska Annaler, 87 A, 2:273–288

Characterizing the Hydrogeology of Bell Harbour Catchment, a Coastal Karstic Aquifer Influenced by the Tide and Affected by a Saltwater Intrusion

M. Perriquet and T. Henry

Abstract A hydrogeological study focused on a karstic aquifer of a small catchment (~50 km^2) located on the south coast of Galway Bay is described. The key aim of the study is to better understand the freshwater/seawater interaction in this coastal region west of Ireland. Discharge from the catchment is entirely through intertidal diffuse springs and submarine groundwater discharges (SGD). Logging of temperature, conductivity and water levels at coastal springs, turloughs and boreholes in the catchment is underway; water samples have been recovered for chemical analysis and water tracing has been undertaken. Initial results clearly show a tidal influence up to 2.5 km inland and an intrusion of seawater at least up to 1 km inland. This saltwater intrusion varies, depending on the balance between the tidal periods (spring/neap) and the groundwater level.

Keywords Coastal karstic aquifer • Saltwater intrusion • Hydrogeology • Catchment discharge • Ireland

1 Introduction

Understanding the interaction of fresh and sea water in coastal karst systems is critical for current and future water management, especially in the context of various EU Directives (Water Framework Directive, Floods Directive, Marine Strategy Framework Directive) and in the context of the potential impacts of changing weather patterns and sea levels. The complex behaviours of groundwater in karst systems highlight

M. Perriquet (✉) · T. Henry
Biogeoscience Group, Earth and Ocean Sciences, School of Natural Sciences, National University of Ireland, Galway, Ireland
e-mail: m.perriquet1@nuigalway.ie

T. Henry
e-mail: tiernan.henry@nuigalway.ie

J. Mudry et al. (eds.), *H2Karst Research in Limestone Hydrogeology*, Environmental Earth Sciences, DOI: 10.1007/978-3-319-06139-9_12,

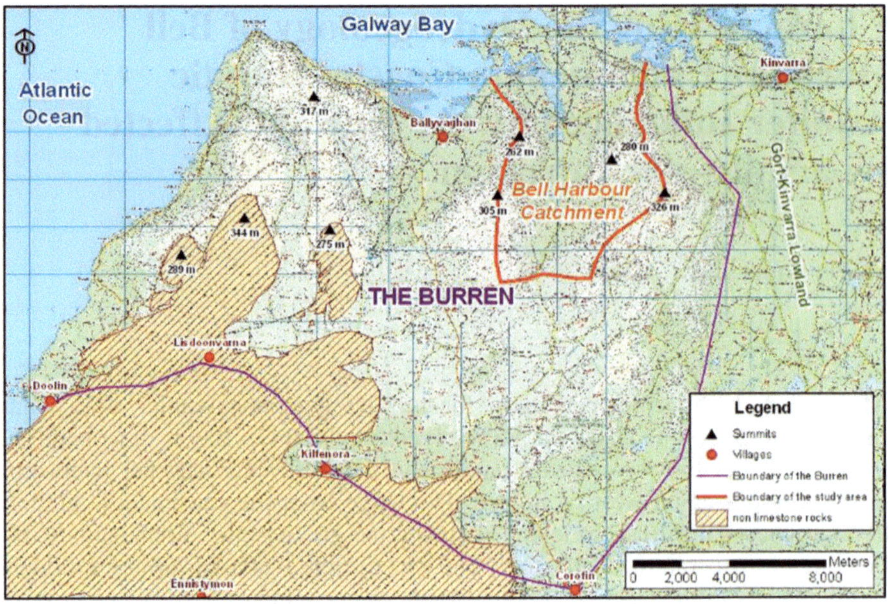

Fig. 1 Location of Bell Harbour catchment

the needs for detailed data collection and analysis at the field scale. Additional complexities areas added in coastal karst areas where sea water and groundwater interact. This study is focused on developing a better understanding of salinity intrusion in a relatively small groundwater body in a large, regional scale karst aquifer system in the west of Ireland. The catchment—Bell Harbour Catchment—was chosen due to its relatively small size and its fairly well-defined boundaries (Drew 1990). It is a sub-catchment of a much larger karst catchment (the Gort-Kinvarra Lowland Catchment).

The study catchment covers an area of around 50 km^2 which is located in a large karstic region called the Burren (Fig. 1) and is defined by upland areas to the west, south and east at an altitude of about 300 m. Drainage is to the north to Galway Bay via Bell Harbour.

The Burren Region is one of the most extensive limestone karst areas in north-western Europe covering an area of approximately 600 km^2 (Gallagher et al. 2006). The geology is dominated by massive or bedded Carboniferous limestone of several hundred metres thickness with interbedded chert, shale and dolomite horizons. The area is bounded to the east by limestones of the Gort Lowlands and to the south by sandstones and shales. Within the study area the geology is dominated by limestones of the Burren and Slievenaglasha Formations, all of which dip gently (2–3°) to the south (Fig. 2). The Burren Formation limestones are generally pale-grey and thickly to massively bedded with occasional cherty intervals (Pracht et al. 2004). The lower boundary of the Burren Formation is marked by a laterally continuous dolomite horizon while the upper boundary is formed by an irregular paleokarst surface. This latter boundary is visible at coastal outcrop and forms the base of the Slievenaglasha

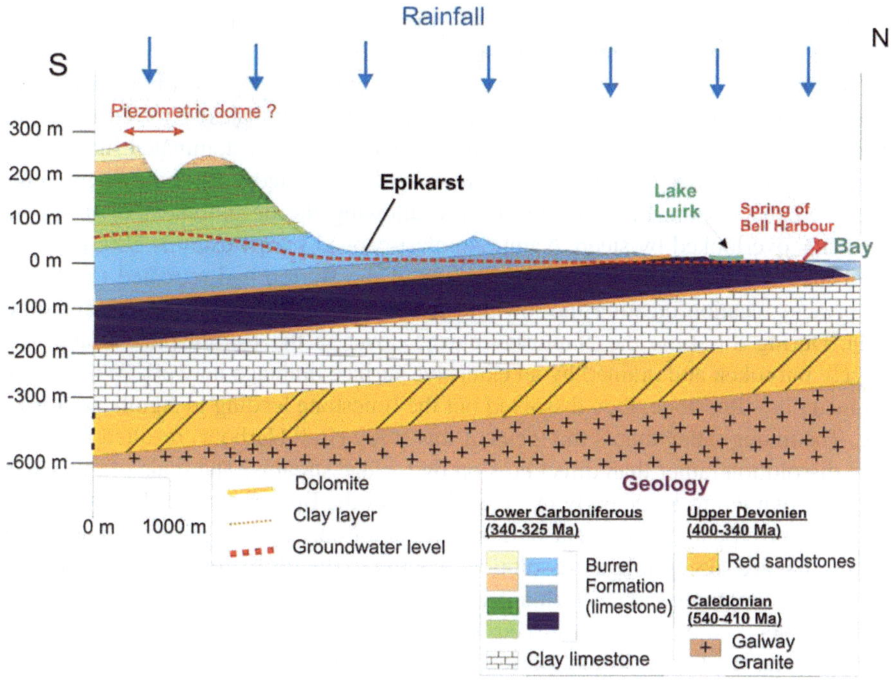

Fig. 2 Geological and hydrogeological profile South–North of Bell Harbour catchment

Formation. The rocks of the Slievenaglasha are predominantly thick-bedded, pale grey crinoidal limestones whose upper boundary (in south County Clare) is marked by a thin phosphatic and cherty micrite and shale (Sleeman and Pracht 1999).

Late Palaeozoic deformation of rocks in Ireland occurred with the continental collision forming Pangea. While the effects of this collision are apparent in the rocks in the southwest of Ireland, the succession in South Galway is largely undisturbed (Graham 2009). The limited deformation of the succession in the north of the Burren has been attributed to the stabilizing presence of a southward continuation of the Galway Granite pluton. The main structures across the entire region are sub-parallel east-west to east-north-easterly trending, gently plunging folds with wavelengths up to 500 m (Pracht et al. 2004). Only two faults are mapped in the Burren region, one of which is in the western part of the current Bell Harbour Catchment study area. This fault, which is known as MacDermott's Fault, runs approximately north-south and shows a slight sinistral displacement of members of the Burren and Slievenaglasha Formations. Geophysical investigation of this fault is being completed, using electrical resistivity tomography, as part of this ongoing research. Initial indications suggest that the fault extends beyond its current mapped extent running under Bell Harbour (O'Connell, Personal Communication 2012). Joints and veins are extensive throughout the area. The veins are clustered, non-stratabound fractures formed in response to north-south compression during the Variscan, while the joints occur in stratabound fracture networks formed during uplift (Gillespie et al. 2001).

By the early Cenozoic, an extensive and thick cover of Upper Carboniferous (and possibly younger) clastic sediments (sandstone, silts and shales) covered the entire region. Under the temperate, humid climatic conditions that dominated through the Cenozoic in Ireland limestone dissolution and associated surface lowering rates exceeded the erosion rates of the clastics, allowing the development of limestone lowlands overlooked by steep, Namurian clast-capped scarps (Simms 2003). By the Pliocene, it is likely that the northern scarp of the Burren had retreated to roughly its present location creating the lowland landscape to the east of the present study area. During this time the Namurian clastic cover in the Burren plateau area was largely unbroken and drained by an extensive surface drainage network. Where these drained onto limestone, they dissolved out the limestone leading to significant lowering. The major valley that is the focus of this study (Bell Harbour) resulted from surface dissolution rather than direct erosion by surface streams. The scale of this valley suggests that it developed early in the evolution of the Burren plateau (Simms 2003).

The Burren and Slievenaglasha Formations are largely composed of pure limestones that are susceptible to dissolution, leading to the formation of distinctive karst topographic features such as turloughs (seasonal lakes largely fed by and draining to groundwater), swallow holes, sinking streams, limestone pavement, caves and large springs. The landscape in the Burren region (and in the study area) is among the best examples of karst landscape in Europe and is the finest example of karst terrain in Ireland (Drew 2001). There are thought to be three main groundwater flow systems operating in this region: (1) rapidly draining epikarstic (shallow) systems made up of solution conduits and joints; (2) deeper discrete conduit flow systems where flow is through major dissolution conduits and fracture conduits; and (3) dispersed fissure/joint systems linked to the larger conduit systems. In the study area there is little evidence of surface drainage features and soil cover is thin or absent in the upper portions of the catchment; drainage is almost entirely by groundwater flow.

Annual precipitation in the area averages approximately 1,500 mm with an annual effective rainfall estimated at 980 mm. The rainfall occurs throughout the year though the spring and early summer months tend to be drier. Recharge in the catchment is mostly diffuse through the large area of limestone pavement in the upper portion of the catchment, and, as is typical for the region, water drains almost wholly by underground channels directly to the sea via submarine or littoral diffuse springs. Aquifer storage is low and during extended wet periods the conduit systems can back up leading to the development of three groundwater-fed and draining seasonal lakes (turloughs) which may persist for weeks or months on the landscape. Five intertidal springs have been located in the eastern edge of Bell Harbour (Fig. 3). The lower portions of the extensive underground drainage system may be affected by saltwater intrusion especially during high tides and or following extended dry periods when groundwater discharge is limited. Moreover, this salinity intrusion can be influenced spatially by the geology and the topography of the catchment as well as by the location, size and shape of the conduits/fractures which carry the groundwater flows.

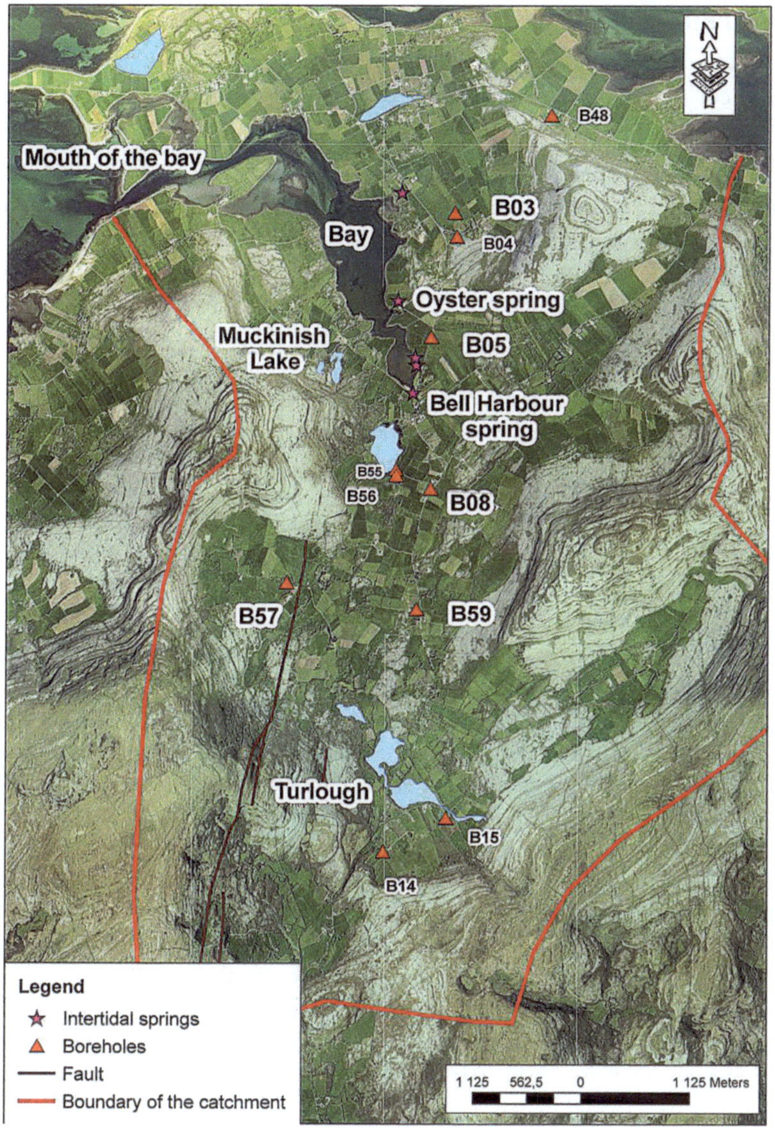

Fig. 3 Location of the hydrogeological elements into Bell Harbour catchment

The research focuses mainly on understanding the relationship between tidal level and groundwater chemistry, using groundwater level and specific conductivity values collected in boreholes, turloughs, springs and in the bay. Time series analysis will be used, among other tools, in the aim to define the tidal influence into the catchment over time. Moreover, a method to evaluate the quantity of freshwater discharging into the Bell Harbour bay from submarine and intertidal groundwater discharge (SiGD) is applied.

2 Materials and Methods

This project is the first detailed groundwater study to be undertaken in this catchment and detailed data collections have been completed on the field and are ongoing:

- A karst geomorphological investigation is in progress, the focus of which is mapping the most important land karst features of the area;
- Water samples from thirteen (13) boreholes, three (3) land springs, seven (7) marine springs, one (1) cave and two (2) turloughs were recovered in March 2011 for chemical analysis (analysis undertaken focused on the abundance of major anions and cations);
- Discrete discharge measurements of the five coastal springs have been recorded on a monthly basis over a one-year period;
- Discrete water level measurements using a dip meter have been recorded in nine boreholes and specific conductivity (SpC) values have been collected at five intertidal springs on a monthly basis over a one-year period.
- In Situ *Aqua TROLL 200* loggers have been installed in five boreholes (B03, B05, B08, B57 and B59) and CTD-Divers are sited at three permanent or seasonal lakes and two coastal springs. The in situ loggers are connected with a vented cable to take account for atmospheric pressure. This sensor can resolve pressure level to ±0.05 % Full Scale, temperature to ±0.01 °C and SpC to ±0.1 μS/cm. A Baro-Diver installed adjacent to borehole B05 allows for compensation of the pressure measurements from the CTD-Divers. These sensors can resolve pressure level to 1 cm H_2O, temperature to ±0.01 °C and SpC to ±0.1 μS/cm. Two SBE 37-SI MicroCats have been ballasted at the bottom of the turloughs and two others are hung on mussel float lines in Bell Harbour bay at 1 and 6 m below the water surface. Pressure readings are not taken for Bell Harbour MicroCats because of tidal movement and fluctuation. These sensors can resolve pressure level to ±0.1 m, temperature to ±0.0001 °C and SpC to ±0.1 μS/cm. The logging interval is 15 min for all of the sensors. An RTK GPS survey of all wells, springs and collection points has been completed to relate all water levels collected above Mean Sea Level (MSL).

3 Results

The collection of the data is still on-going but already early results are presented here, and can be grouped into three main areas:

1. Data from the various loggers are used to understand the principal characteristics of the aquifer and to explore the relationship between rainfall events and well water levels;
2. All of the data sets are used to better understand the influence of the tide on saltwater intrusion; and,

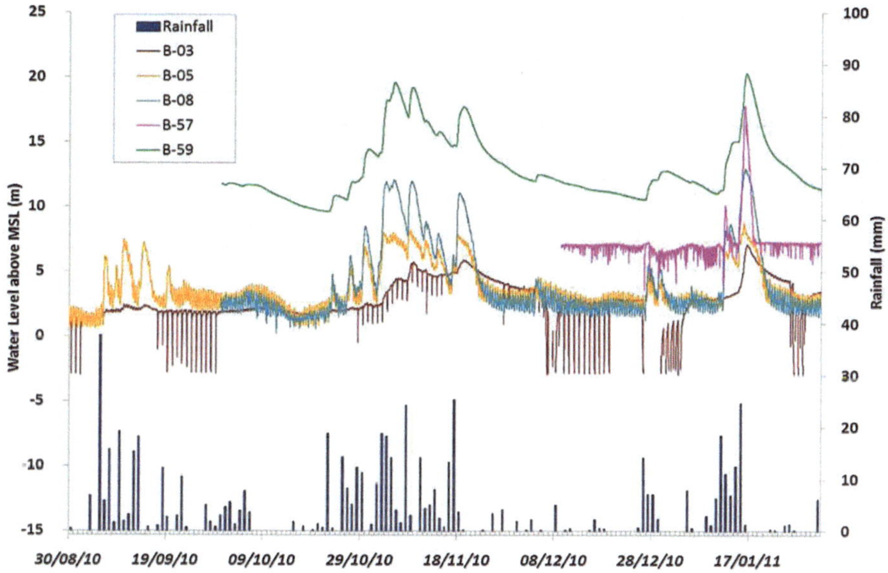

Fig. 4 Water levels recorded in five boreholes plotted with rainfall (NUI Galway) from the 30th of August 2010 to the 31th of January 2011

3. Estimation of the submarine and intertidal groundwater discharge into the bay has been attempted to better understand temporal fluctuations in groundwater quantities.

3.1 Aquifer Water Level Changes

Water level measurements in the five boreholes logged show a range of responses (Fig. 4). During a period of base flow, boreholes B03, B05 and B08, located within 1 km of the sea, have almost the same elevation (~2 m above MSL). The base-flow elevations from boreholes B57 and B59 (~2.4 km from the shore in the middle of the valley) are 7 and 10 m above MSL, respectively.

The responses after a rainfall event are shown in Table 1. Borehole B03 is completed in the solid limestone matrix (or in fracture system not well connected with an active conduit), and it shows slow storm responses and only small changes in water levels for the two rainfall events. It is also affected periodically by pumping effects with a large drawdown of 5 m. B05 shows rapid storm responses and fairly large rises in levels which suggests it is well connected to an active conduit system and that recharge reaches it rapidly. B08, B57 and B59 display amplitudes of 10 m each: suggesting that a large volume of groundwater discharges through the valley. B57 responds quickly, suggesting it is near a conduit system, probably the large mapped NNE-SSW trending fault (MacDermott's Fault).

Table 1 Delay and amplitude of the water level changes in the five boreholes between the first rainfall input and the maximum of water level rising for two major rainfall events

Rainfall event	October 31, 2010–November 5, 2010		January 10–15, 2011	
Intensity of rainfall	11 mm/day		13 mm/day	
Boreholes	Amplitude (m)	Delay (h)	Amplitude (m)	Delay (h)
B03	~4	~48	~4	~48
B05	~6	~18	~6.5	~24
B08	~10	~35	~10.5	~35
B57	N/A	N/A	~10.5	~18
B59	~10	~43	~9	~41

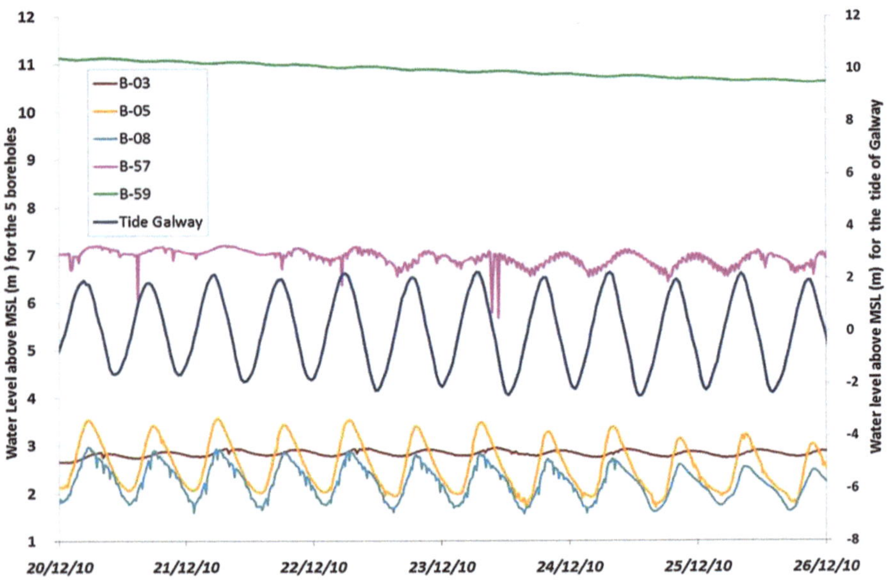

Fig. 5 Tidal influences observed in water levels recorded in five boreholes from the 20th to the 26th of December 2010

3.2 Tidal Influence and Seawater Intrusion

Water levels recorded in the five boreholes are all tidally influenced (Fig. 5). Amplitudes vary from 0.04 m for water levels at B59 to 1.4 m for water levels at B05. Tidal efficiency—the ratio between the amplitude observed at a borehole and the amplitude of the tide in the bay—values were calculated (Table 2). The lag time observed with the Galway tide is from 1 h 35 min at Borehole B05 to 8 h and 45 min at B59. Small variations observed in water levels of boreholes B05, B08 and especially in B57 are only due to pumping effects.

Examination of borehole B05 SpC data shows the largest variations associated with tidal influence (Fig. 6). Fluctuations of SpC are stronger during spring tides

Table 2 Distance from the shore, average tidal efficiency and lag time estimated for the 5 boreholes

	Distance from the shore (m)	Tidal efficiency (m)	Lag time
B03	560	0.058	3 h 30 min
B05	265	0.445	1 h 35 min
B08	1,065	0.174	2 h 00 min
B57	2,400	0.131	3 h 35 min
B59	2,300	0.005	8 h 45 min

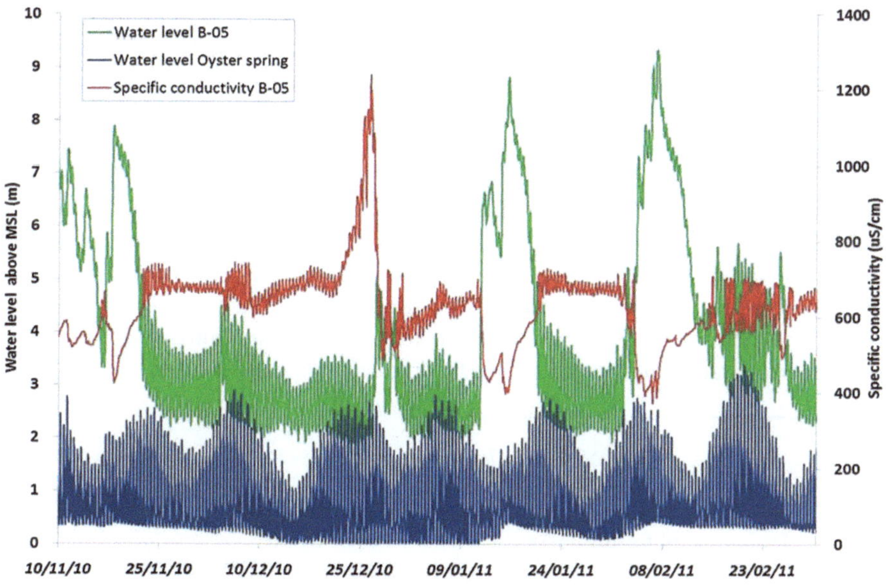

Fig. 6 Water levels recorded at Oyster spring with water levels and SpC of Borehole B-05 from 10/11/2010 to 02/02/2011

(generally varies between 60 and 110 μS/cm) than during neap tides (vary between 10 and 40 μS/cm). During base-flow periods the SpC is close to 650 μS/cm and can drop to 380 μS/cm during a high rainfall event (during which the water level rises about 6.5 m). A high peak of SpC (1227 μs/cm) was observed on December 26, 2010. This occurred during a spring tide and when the Oyster spring was completely dry (from December 13, 2010 to January 10, 2011) and precipitation was very low (<50 mm from December 12, 2010 to January 9, 2011). The water level was the lowest recorded (1.7 m above MSL at low tide) since August 2010 during this same period. This high peak of SpC can be associated with a seawater intrusion into the aquifer which occurred during a period of low rainfall and high tidal amplitude.

Only the SpC data recorded at borehole B08 has similar tidal variations with smaller amplitude. However, no saltwater intrusion is observed into this borehole over the time period. The other boreholes are either further inland (B57, B59) or not well connected with the conduit/fracture system (B03) and their SpC data does not show any fluctuations due to the tide or a saltwater intrusion.

3.3 Submarine and Intertidal Groundwater Discharge

An estimation of the groundwater volume discharging into Bell Harbour was made using the simple tidal prism model. This method is explained in detail in Cave and Henry (2011): a volume of water available for runoff and a volume of freshwater discharging into the bay are estimated separately and then compared.

The salinity of water between each high to low tide is averaged and divided by the maximum salinity measured on the previous flood tide (Eq. 1). This allows for some ebb water being returned to the bay on the following flood tide. The height of low water is subtracted from the height of high water for that tide, and the result multiplied by the surface area of the estuary, to obtain the total volume of water brought out on the ebb tide. The total ebb tide volume is then multiplied by the freshwater proportion to obtain the volume of fresh water removed on each ebb tide (Eq. 1). It is assumed that the water in the bay is well mixed, and that the Microcat salinity values are therefore representative of the entire thickness of the water column.

$$\overbrace{1 - (S_{HW} - S_{LW})/max.\ salinity\ of\ flood\ tide}^{\text{Proportion of freshwater}} \times \underbrace{(H_{HW} - H_{LW}) \times Surface\ Area}_{\text{Volume of water brought out on the ebb tide}} \quad (1)$$

where S_{HW}–S_{LW} and H_{HW}-H_{LW} are, respectively, the average of salinity and the difference of height between high tide and low tide.

The surface area of Bell Harbour has been calculated from LIDAR data to be 1.8 km^2. This formulation was applied for each tidal cycle from the December 2, 2010 to May 31, 2011 (using the salinity data available from the Microcat located in the middle of the bay).

The Potential Runoff is estimated by the multiplication of the Effective rainfall (ER) (Precipitation—Evaporation) with the surface area of the catchment. The storage term is neglected here due to a regular rainfall through the year. ER was calculated by averaging the daily rainfall from existing Met Éireann rain gauges at Ballyvaughan (located west of the catchment near the coast) and Carran (located inland south-east of the catchment) and subtracting the monthly evaporation data from the station at Birr, the closest to the catchment.

A total freshwater discharge into the Bay of $\mathbf{7.40 \times 10^7\ m^3}$ has been estimated for this 6-month period with the proportion of freshwater averaged at 3.7 % of total water in the bay, giving a freshwater discharge of 4.7 m^3/s (Fig. 7). The total volume of the potential runoff has been estimated at $\mathbf{4.70 \times 10^7\ m^3}$ which is significantly

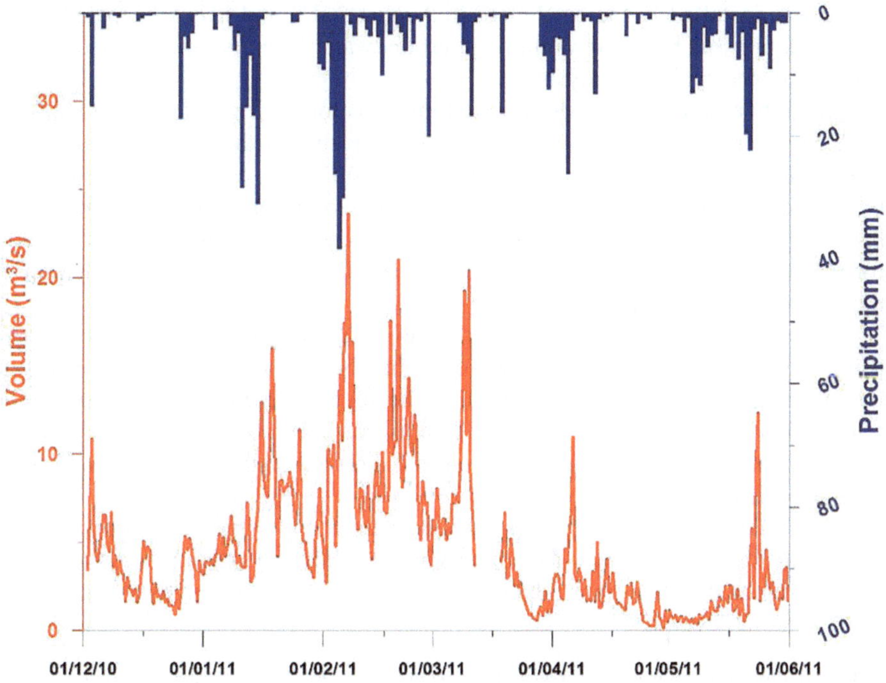

Fig. 7 Estimation of the volume of freshwater discharging into the bay plotted with the volume of the total rainfall

smaller. It appears that the magnitude of the freshwater discharge is over-estimated. However, this method assumes that the water is well mixed throughout the bay; however, Bell Harbour bay behaves differently. Indeed, several dark spots into the bay have been noticed from aerial photographs, and a Microcat had been left just above one of these which could be a potential SGD. To check the nature of these possible SGD locations, a horizontal and vertical salinity survey was completed across the bay on a rainy day in June 2012. Initial results show a drop of the salinity at the spot where the logger had been left. It may therefore be possible that the salinity measured at this point is under-representative of the salinity of the entire bay.

4 Conclusions

The study of a karst aquifer affected by a saltwater intrusion is complex and requires large amounts of data from different sources to understand the dynamic of the saltwater-freshwater interface.

The interpretation of the water level variations in the aquifer shows some differences in the spatial behaviour of the aquifer system and can be used to assess how well or poorly connected each well is to a conduit/fracture system. A rain gauge was installed in the catchment in August 2012, and it is anticipated that a better relationship between rainfall and groundwater level may be estimated using correlation and spectral analysis.

All the boreholes logged show a tidal influence on their groundwater level with different lag times and amplitudes. Using these results with the Ferris and Branch (1952) equations, it is planned to evaluate a transmissivity value at different points in the aquifer. This method will also be used to better understand the spatial behaviour of the aquifer. It is also hoped to examine the data for different periods, for example, during neap or spring tide, or low and high water level to determine the nature of their impacts.

The estimated SiGD volume entering Bell Harbour has been over-estimated using the tidal prism method. However, it has been noticed this over-evaluation might be due to the specific location of the logger which was close to a SGD point. A horizontal stratification of the salinity has been observed during the recent survey work, and it will be incorporated into a revised estimate of freshwater discharge to the bay.

The initial results presented here will be refined and subject to reworking coupled with new data from water geochemical analyses, ongoing monitoring of water levels and SpC in intertidal springs, boreholes and turloughs (which reflect local water tables).

Acknowledgments Based on a research grant aided by the Department of Communications, Energy and Natural Resources under the National Geoscience Programme 2007–2013.

References

Cave RR, Henry T (2011) Intertidal and submarine groundwater discharge on the west coast of Ireland. Estuar Coast Shelf Sci 92(3):415–423

Drew D (1990) The hydrology of the Burren county Clare. Irish Geogr. 23:69–89

Drew D (2001) Classic landforms of the Burren karst. Geographical Association in conjunction with the British Geomorphological Research Group

Ferris J, Branch GSGW (1952) Cyclic fluctuations of water level as a basis for determining aquifer transmissibility. US Dept. of the Interior, Geological Survey, Water Resources Division, Ground Water Branch

Gallagher SJ, MacDermot CV, Somerville ID, Pracht M, Sleeman AG (2006) Biostratigraphy, microfacies and depositional environments of upper Visean limestones from the Burren region, County Clare, Ireland. Geol J 41:61–91

Gillespie PA, Walsh JJ, Watterson J, Bonson CG, Manzocchi T (2001) Scaling relationships of joint and vein arrays from the Burren, co., Clare Ireland. J. Struct Geol 23:183–201

Graham J (2009) Variscan Deformation and Metamorphism. In: Holland CH, Sanders IS (eds) The Geology of Ireland, 2nd edn. Dunedin Academic Press, Edinburgh

Pracht M, Lees A, Leake B, Feely M, Long B, Morris J, McConnell B (2004) Geology of Galway Bay: A geological description to accompany the Bedrock Geology 1:100,000 Scale Map Series, Sheet 14, Galway Bay. Geological Survey of Ireland

Simms M (2003) The Geology of the Burren and the Gort lowlands: In: Mullan G (ed). Caves of County Clare and South Galway. University of Bristol Speleological Societ

Sleeman AG, Pracht M (1999) Geology of the Shannon Estuary: a geological description of the Shannon Estuary region including parts of Clare, Limerick and Kerry, to accompany the Bedrock Geology 1:100,000 Scale Map Series, Sheet 17, Shannon Estuary. Geological Survey of Ireland

Contaminant Attenuation in Karst Aquifers: A Paradigm Shift

M. Sinreich

Abstract Significant advances have been made in the characterization of transport and storage in karst aquifers over recent decades. This improved understanding permits further integration of the behaviour of individual contaminants and their specific transport, enabling comparisons to be made. This has been particularly challenging as it is necessary to consider different flow components encountered in karst aquifers, including fast conduit flow and storage in less permeable rock volumes. Comparative tracing experiments using contaminant surrogates have proved to be an appropriate method for estimating the specific attenuation of selected substances at the field scale. Several attenuation processes may be involved and could be identified. Examples from Swiss karst aquifers highlight the in situ effectiveness of such attenuation processes, some of which can be described using first-order kinetics. It could be shown that solute and colloid tracers are able to interact with aquifer material despite the dominance of preferential and conduit flow components. Consequently, if reactive and/or non-persistent contaminants are involved, the arrival at karst springs is determined by contaminant-specific properties and hydrochemical characteristics rather than by the intrinsic vulnerability of the aquifer. This demands more refined conceptual transport models and also represents a paradigm shift in the assessment of karst groundwater vulnerability and contaminant attenuation.

Keywords Karst aquifers • Tracing experiments • Contaminant vulnerability • Attenuation processes

M. Sinreich (✉)
Swiss Federal Office for the Environment (FOEN),
Hydrogeology Section, 3003 Bern, Switzerland
e-mail: michael.sinreich@bafu.admin.ch

J. Mudry et al. (eds.), *H2Karst Research in Limestone Hydrogeology*,
Environmental Earth Sciences, DOI: 10.1007/978-3-319-06139-9_13,

1 Introduction

One of the most cited characteristics of karst aquifers is their particular vulnerability to contamination, which is linked to fast water flow along conduits, as well as the reduced effectiveness of natural attenuation processes compared to other aquifer types. The high vulnerability of karst groundwater is evidenced by numerous studies dealing with the presence of many types of contaminant found at karst springs once released in the catchment.

Transport, storage and attenuation are key factors which determine the fate of contaminants in groundwater and their appearance at monitoring points. While karst studies over recent decades have led to a better understanding of the hydrodynamics of such systems and have produced updated conceptual models of transport and storage, investigation of attenuation processes seems to have lagged behind.

Although contaminant-specific vulnerability has been addressed in relation to karst groundwater management on a conceptual basis (Zwahlen 2004), research is still required to provide more information on mechanisms affecting the fate and transport of specific contaminant types in karst aquifers. The actual paradigm regarding karst vulnerability features both preferential conduit flow and the bypassing of protective layers; however, it is still subject to underestimation of attenuation potential, as significant interaction between contaminants and rock surfaces along these passages is neglected.

This paper intends to focus on the potential of karst aquifers to attenuate reactive and/or degradable contaminants during transport and to attempt quantification of such phenomena on a process-scale. The findings presented suggest a paradigm shift toward the consideration of attenuation processes and their integration into karst groundwater transport conceptual models.

2 Contaminant-Specific Vulnerability

Concepts dealing with karst groundwater vulnerability in Zwahlen (2004) consider transit time and maximum concentration as critical parameters for defining vulnerability to contamination. For intrinsic vulnerability, contaminant concentration is a function of attenuation, which, in turn, is the result of dilution during advective-dispersive transport. In the present context, however, attenuation rather refers to reaction processes related to specific vulnerability, defined as an additional effect dependent on an individual contaminant or contaminant type. According to Fig. 1, which shows possible classes of karst groundwater vulnerability (Sinreich et al. 2007), consideration of contaminant-specific attenuation processes may lead to a reduced vulnerability.

Dual porosity and multi-permeability approaches are possible hydrodynamic conceptual models that may form the basis for specific attenuation assessment (Fig. 2). Although a fast flow component must always be expected in developed karst systems (resulting in short first arrival times), slow flow components through

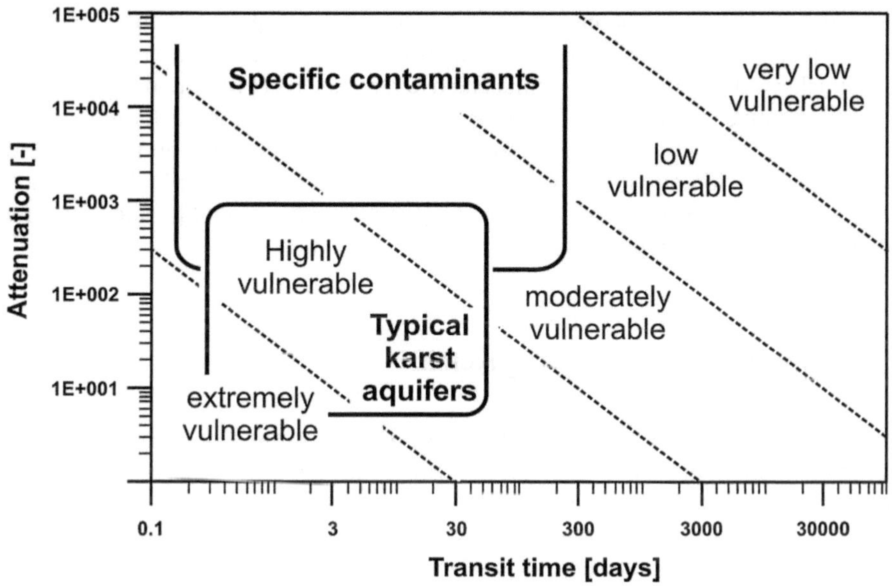

Fig. 1 Vulnerability classification with reduced vulnerability when additional attenuation is provided by contaminant-specific processes

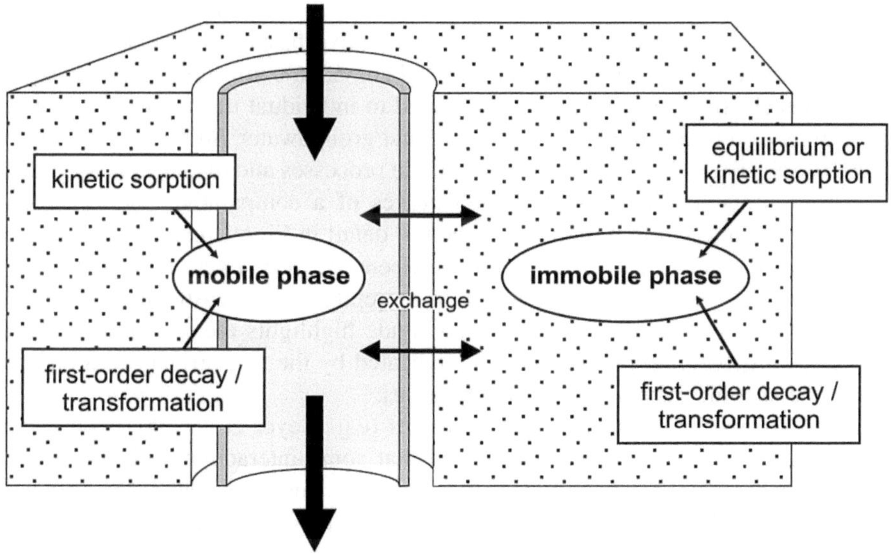

Fig. 2 Conceptual model for a karst sub-system including a fast flow component and a less permeable rock volume, and assuming characteristic attenuation reactions and kinetics in the different domains (Sinreich et al. 2007)

less permeable rock volumes, including storage, also play an important role. Consequently, discrimination of sub-systems is essential to evaluate their attenuation potential, assuming distinct hydraulic and hydrogeochemical conditions within each.

For instance, significant water volumes may be put into storage during recharge events which slowly discharge into the karst conduit network during baseflow. This may provide significantly longer residence times during which substances are able to undergo reactions. Storage may thus become a multiplicator for time-dependent attenuation processes.

3 Comparative Tracing Approach

Although groundwater quality monitoring at karst springs is a crucial tool for characterizing contaminant transport and its relation to discharge fluctuations, it often remains qualitative, i.e., no relation to source concentrations. This makes it particularly difficult to obtain recovery rates or attenuation rates from such data. Contaminant monitoring by passive sampling can only become quantitative if combined with an input/output mass balance, as performed by Schwarz et al. (2011) for the transport of polycyclic aromatic hydrocarbons (PAH). Such comprehensive studies are overdue for karst systems though are still rare due to the significant effort necessary to tackle heterogeneity and transient conditions inherent in this aquifer type.

Alternatively, artificial tracer testing can be used to provide a means for determining transport parameters on a more quantitative basis. While classical tracer tests are limited to tracers with conservative behaviour, comparative tracing experiments benefit from specific processes related to individual tracers, which may act as surrogates for contaminants relevant to karst groundwater. Such experiments are more conclusive if they are restricted to single processes and distinct sub-systems.

Results obtained from breakthrough curves of a comparative tracing experiment, as presented in Fig. 3 and described in detail in Sinreich and Flynn (2011), emphasize the remarkable distinction between recovery rates, and attenuation, respectively, of various substances within the epikarst layer. Attenuation extending to over more than two orders of magnitude highlights the relevance of some processes involved, despite fast flow as indicated by the first arrival times and the steep rise of some tracers' breakthrough curves.

Similar trends reported from other studies (e.g. Geyer et al. 2007) prove that tracer-specific processes are involved and that some interaction with the aquifer surfaces may occur. If viewed in isolation, however, such observations are not sufficient to fully characterize attenuation and related processes and have limited significance. Complementary laboratory-scale experiments can solve this problem by identifying the processes that are responsible for tracer attenuation when aquifer material is contacted.

Provided that advective-dispersive transport can be assumed, tracer-specific attenuation can be directly inferred from the divergence in conservative solute and

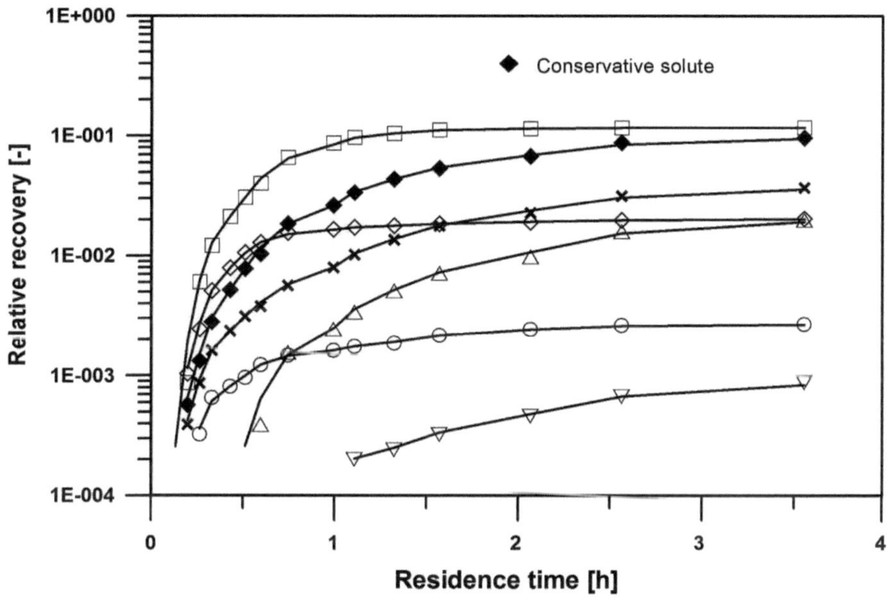

Fig. 3 Cumulative mass recovery during a comparative tracing experiment indicating significant differences for inorganic solutes, organic solutes as well as biotic and abiotic colloids due to specific attenuation of each tracer

reactive solute breakthrough curves (Sinreich and Kozel 2009). Similarly, particle attenuation can be deduced if a non-reactive reference particle tracer of identical size is injected (Sinreich et al. 2009).

4 Process-Based Assessment

In the following, some experiments are briefly described in order to evaluate the potential of solute and particulate substances to undergo reactions with karst aquifer material inferred from field tracing in conjunction with information obtained from laboratory experiments. The aim is to help gain an insight into processes at work during transport and to highlight the role that both layer and contaminant characteristics, as well as hydrochemical conditions and variations, can play in this context.

4.1 Layer Characteristics and Reactivity

Layer properties are determined by mineralogical composition as well as microbial activity in the subsurface. The role of biofilms on contaminant fate is manifold and extends from degradation of organic compounds to providing specific sorption

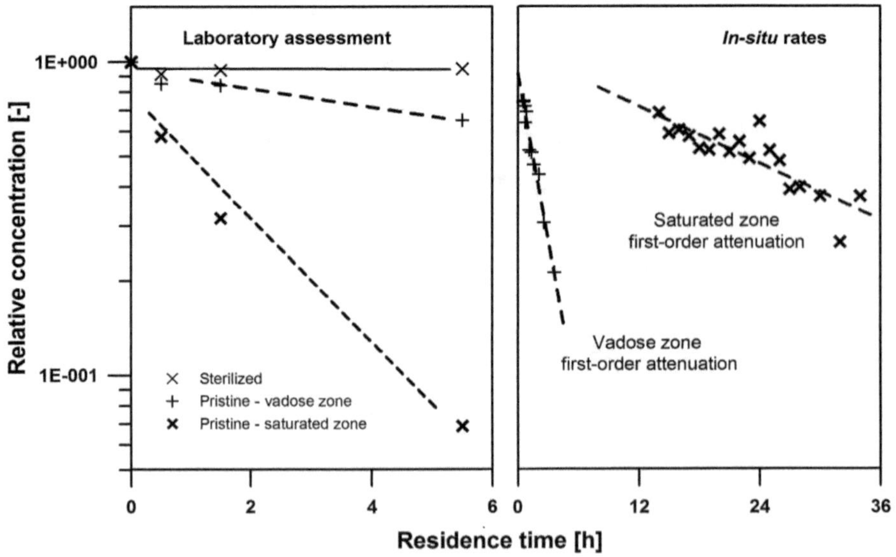

Fig. 4 Solute biodegradation in microcosms of karst aquifer materials (*left*) and in situ attenuation rates deduced from tracer breakthrough curves assuming first-order kinetics (*right*)

sites by altering rock surfaces (Sinreich and Kozel 2009). The example of Fig. 4 highlights the importance of different layer characteristics on solute attenuation, including residence time within each sub-system

Experiments shown in Fig. 4 investigated and quantified mass removal from karst aquifers caused by degradation processes (Sinreich 2007). The results of laboratory batch tests that compared pristine and sterilized aquifer microcosms suggest that biofilms are responsible for the attenuation of a non-persistent solute tracer. It concludes that biodegradation is the relevant process responsible for tracer attenuation under the conditions encountered.

In situ rates deduced from breakthrough curves of a conservative tracer and the non-persistent tracer for two field tracing experiments exhibit degradation rate in the saturated zone which was one order of magnitude lower than in the vadose zone. Given the much longer travel time during saturated flow transport in that case, mass loss is about 50 % for both sub-systems during the experiments.

4.2 Contaminant Characteristics and Reactivity

Results from a vadose zone experiment emphasize the role of contaminant properties on attenuation, employing microorganism types of the same size but with differing sorption characteristics.

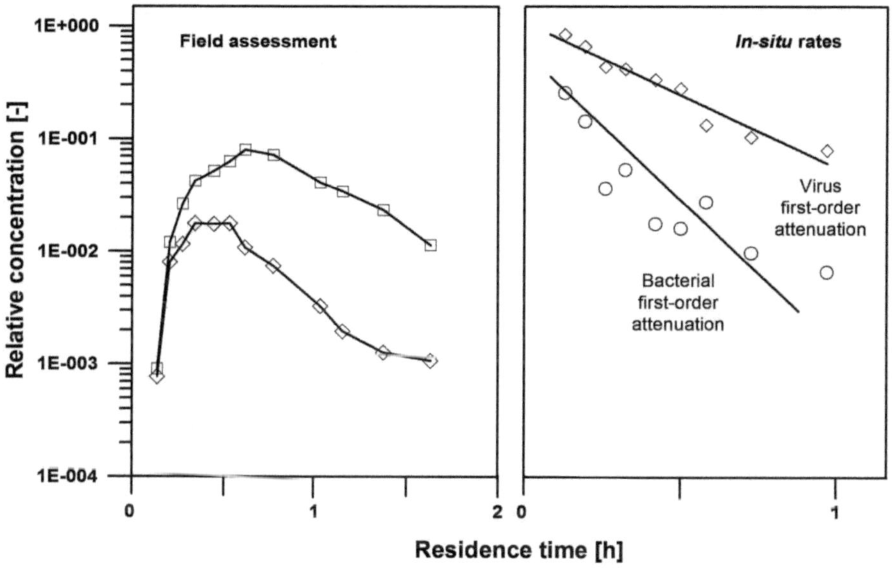

Fig. 5 Breakthrough curves of two virus types indicating significant attenuation of the reactive type (*left*). The deduced first-order sorption rate is compared to that of a similarly assessed bacteria-sized colloid tracer (*right*)

Figure 5 reveals the presence of sorption processes, resulting in differences in recovery and concentration of similarly-sized colloids which can exceed two orders of magnitude. Laboratory tests proved that biotic processes are not involved during the timeframe of the field experiments and that sorption processes must be responsible for the removal of particulates from suspension. Consequently, significant in situ attenuation for both bacteria-sized and virus-sized colloids must be assumed, due to strong interaction with, and attachment to, rock surfaces. This is particularly interesting, as colloids have been shown travelling predominantly along the fast flow component due to exclusion effects, thereby by-passing the less permeable rock volume (Sinreich et al. 2009).

4.3 Hydrochemical Variations

Physico-chemical reactions are strongly dependent on the ambient conditions encountered in the system, such as pH, redox, or ionic strength (White 1997). These conditions may fluctuate widely in karst aquifers, both spatially and temporally.

Figure 6 shows another example of tracing using the same reactive virus type as in Fig. 5, in order to demonstrate the effect of hydrochemical variations on virus attenuation over time. Due to the simultaneous injection of ionic tracers during

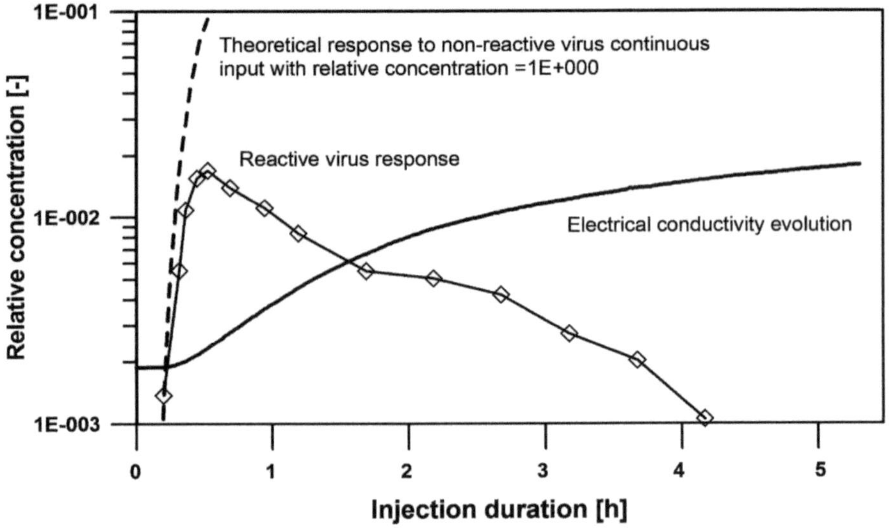

Fig. 6 Gradual decline in virus breakthrough during continuous injection due to an increasing attenuation rate along with changing hydrochemical conditions

the continuous input experiment, the evolution of virus concentration mirrors the gradual rise of ionic charge in groundwater, expressed by the increasing electrical conductivity.

A tracer concentration which increases towards unity could be expected for a non-reactive colloid. Assuming the same virus attenuation rate as for Fig. 5, a constant relative concentration of about 0.12 could be modeled (Flynn and Sinreich 2010). However, as with the change in hydrochemical conditions, the decline in virus concentration continues until virtually no viruses reach the monitoring point, despite continuous injection of this tracer. This in turn suggests that all colloids are able to have contact with rock surfaces and that attenuation is controlled by virus properties and hydrochemistry, and is not restricted by limited possibilities for interactions. Modelling results accordingly revealed increasing attenuation rates with time.

5 Concepts and Conclusions

The current understanding of karst hydrogeology forms a good basis for the refinement of according conceptual models and expansion to include reactive transport phenomena which consider contaminant-specific processes. Some promising approaches have been proposed by several authors, such as Loop and White (2001) as well as Vesper (2008) for NAPL, Savoy (2008) for degradable compounds, or Schwarz et al. (2011) for PAH. Experimental data from comparative

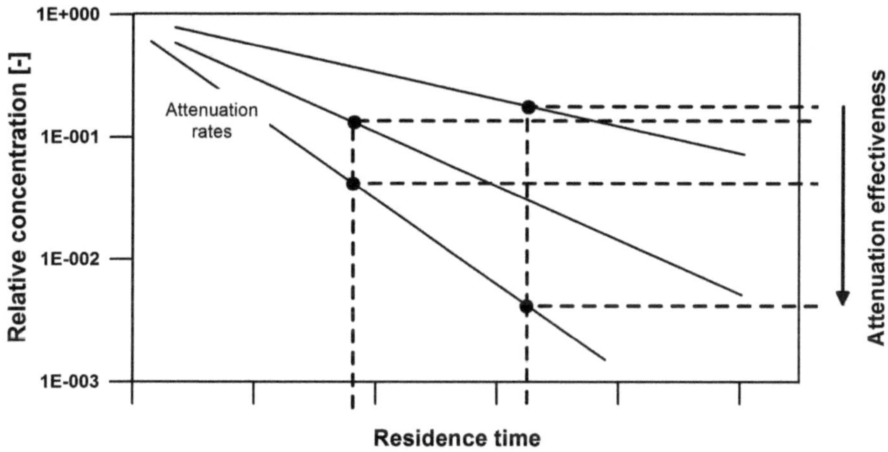

Fig. 7 Conceptual intersection of attenuation rates, depending on contaminant and layer characteristics, with residence times encountered in karst systems, which yields in situ attenuation effectiveness

tracing is assumed to provide significant input in such developments (Sinreich 2011). Supplemental tools, such as isotope fractionation or modelling approaches, have the potential to make further progress in this field.

In specific terms, the findings have proven the existence and time-dependent nature of several attenuation processes in karst sub-systems. As can be seen in the examples of Figs. 4 and 5, process modelling, which assumes first-order kinetics, can explain the behaviour of different contaminant types and therefore represents an appropriate concept in attenuation assessment. Although laboratory experiments can identify the relevant processes which contaminants or tracer surrogates undergo in the presence of aquifer material, field-based tracer testing has the potential to assess the in situ effectiveness of such processes.

These findings contradict the common hypothesis of the insufficient opportunity for contact of contaminants with reactive surfaces in the karst subsurface. In fact, attenuation limitation is suggested to be primarily the result of the lack of reactivity of some contaminants. If, however, contaminants are prone to attenuation processes, karst environments are capable of providing a significant attenuation potential. In conclusion, provided there is a certain reactivity of the contaminant of interest, attenuation processes must be assumed. This applies to both solutes and particles as well as for vadose and saturated karst sub-systems. Attenuation effectiveness is then determined by reaction kinetics in conjunction with residence time (Fig. 7).

The time-dependency of most processes, therefore, make them less efficient in karst than in other aquifer types. However, the results also indicate that reaction kinetics can be very rapid and be able to totally remove contaminants from the liquid phase within hours. This timeframe makes the processes relevant for karst groundwater flow conditions. Furthermore, prolonged residence times are a key factor for the effectiveness of attenuation processes in karst aquifers, as is the case with storage volumes or flow through low permeability zones.

The evidence for the in situ effectiveness of attenuation processes, even in karst, despite its relatively unfavourable conditions, suggest that the hypothesis of the overall high vulnerability of karst aquifers should be relativized with respect to contaminant-specific fate and transport in the aquifer. This represents a paradigm shift in karst hydrogeology and forces karst groundwater transport vulnerability to be assessed contaminant-specifically, and on a process-scale.

References

Flynn R, Sinreich M (2010) Characterisation of virus transport and attenuation in epikarst using short pulse and prolonged injection multi-tracer testing. Water Res 44(4):1138–1149

Geyer T, Birk S, Licha T, Liedl R, Sauter M (2007) Multitracer test approach to characterize reactive transport in karst aquifers. Ground Water 45(1):36–45

Loop CM, White WB (2001) A conceptual model for DNAPL transport in karst ground water basins. Ground Water 39(1):119–127

Savoy L (2008) Use of natural and artificial reactive tracers to investigate the transfer of solutes in karst systems. PhD thesis, University of Neuchâtel, 194 pp

Schwarz K, Gocht T, Grathwohl P (2011) Transport of polycyclic aromatic hydrocarbons in highly vulnerable karst systems. Environ Pollut 159(1):133–139

Sinreich M (2007) Biodegradation potential in karst settings indicated by the pyranine dye tracer. In: 35th IAH Congress, groundwater and ecosystems, Lisbon, pp 308-309

Sinreich M (2011) Towards developing conceptual models for reactive contaminant transport in karst. IAHS Publ 342:473–476

Sinreich M, Flynn R (2011) Comparative tracing experiments to investigate epikarst structural and compositional heterogeneity. Speleogenesis Evol Karst Aquifers 10:60–67. www.speleogenesis.info

Sinreich M, Kozel R (2009) Karst aquifer pollution—Examining the role of biofilm coatings. In: HydroEco'2009, ecosystems interfacing with groundwater and surface water, Vienna, pp 45–48

Sinreich M, Cornaton F, Zwahlen F (2007) Evaluation of reactive transport parameters to assess specific vulnerability in karst systems. IAH-SP 11:21–31

Sinreich M, Flynn R, Zopfi J (2009) Use of particulate surrogates for assessing microbial mobility in subsurface ecosystems. Hydrogeol J 17(1):49–59

Vesper DJ (2008) Karst resources and other applied issues. In: Martin JB, White WB (eds) Frontiers of karst research. Karst Waters Institute, Leesburg, pp 65–73 (Spec Publ 13)

White WB (1997) Thermodynamic equilibrium, activation barriers, and reaction mechanisms for chemical reactions in Karst Terrains. Environ Geol 30(1–2):46–58

Zwahlen (ed) (2004) Vulnerability and risk mapping for the protection of carbonate (karst) aquifers, COST Action 620. European Commission, Brussels/Luxembourg, 297 pp

MIX
Papier aus verantwortungsvollen Quellen
Paper from responsible sources
FSC® C105338

If you have any concerns about our products,
you can contact us on
ProductSafety@springernature.com

In case Publisher is established outside the EU,
the EU authorized representative is:
Springer Nature Customer Service Center GmbH
Europaplatz 3, 69115 Heidelberg, Germany

Printed by Libri Plureos GmbH
in Hamburg, Germany